LIGHTNING

The Earth series traces the historical significance and cultural history of natural phenomena. Written by experts who are passionate about their subject, titles in the series bring together science, art, literature, mythology, religion and popular culture, exploring and explaining the planet we inhabit in new and exciting ways.

Series editor: Daniel Allen

In the same series
*Air* Peter Adey
*Cave* Ralph Crane and Lisa Fletcher
*Desert* Roslynn D. Haynes
*Earthquake* Andrew Robinson
*Fire* Stephen J. Pyne
*Flood* John Withington
*Islands* Stephen A. Royle
*Lightning* Derek M. Elsom
*Meteorite* Maria Golia
*Moon* Edgar Williams
*Tsunami* Richard Hamblyn
*Volcano* James Hamilton
*Water* Veronica Strang
*Waterfall* Brian J. Hudson

# Lightning

Derek M. Elsom

REAKTION BOOKS

*For Elizabeth, my wife*

Published by
Reaktion Books Ltd
33 Great Sutton Street
London EC1V 0DX, UK
www.reaktionbooks.co.uk

First published 2015

Printed and bound in China

A catalogue record for this book is available from the British Library

ISBN 978 1 78023 496 0

# CONTENTS

# Preface

Thunder is good, thunder is impressive; but it is lightning that does the work.[1]

Thunderstorms produce thunder and lightning. Although thunder can rattle windows, trigger vehicle alarms and terrify some people, it is lightning that causes serious damage and disruption to our lives, even threatening us directly with injury and death. Lightning is awesome and beautiful to watch, but it is dangerous too. With around four million lightning flashes happening every day throughout the world, it is a weather threat that demands our attention and respect.

The content of this book reflects my lifetime's interest in enjoying the spectacle of lightning – nature's own stunning firework display. Each lightning display is unique in form, composition and colour. My fascination with lightning has encouraged a desire to understand how lightning has featured in the lives of people in the past and present in terms of mythology, folklore, science and the arts.

The changing attitudes to lightning are also explored. The belief that lightning was so powerful that only gods and goddesses could generate and control it dominated early civilizations. They created thunder, lightning or simply storm deities in many forms, each with a different means of generating thunder and lightning. By medieval times these storm deities were supplanted by modern religious beliefs, although folklore concerning the old gods and goddesses was retained by some individuals and communities. Consequently, objects associated with the old deities – stone 'thunderbolts', certain plants and other items thought to have magical properties – were used to protect families and their homes from being struck by lightning.

Stunning lightning display from a supercell thunderstorm, Nevada, u.s.

In medieval times the fear that lightning could be directed by evil individuals such as witches resulted in many of those accused being tortured and put to death. Although the persecution of lightning-controlling witches ended in the eighteenth century in Europe and North America, it remains a belief within some traditional communities in countries across the world. In 1990 South Africa even established ten remote villages as safe havens for hundreds of people accused of practising witchcraft.

Since the mid-eighteenth century, science has helped explain the nature and formation of lightning. Proof that lightning is an electric current (or 'electric fluid' as it was initially called in popular terms) has enabled us to develop ways to minimize the damage and disruption it causes. The process by which lightning is formed is explained in this book along with the actions being taken to tackle the threat lightning poses to our forests, buildings, power supplies, aircraft and spacecraft.

The risks of people being struck and injured by lightning are explored and the ways to minimize the chances of being struck – places to avoid and areas to seek shelter, both outdoors and indoors – are discussed. This will help readers avoid being in the wrong place at the wrong time when thunderstorms develop.

Finally, the many ways in which lightning pervades art, literature and popular culture are outlined, with examples given of how lightning features in everyday expressions, books, films, comic strips, paintings, heraldry and sculptures.

This book is wide-ranging in its coverage of lightning-related subjects and, for those readers wanting to find out even more about a specific topic, I hope it marks the beginning of many more years of adding to their knowledge and appreciation of lightning.

# 1 Weapon of the Gods and Goddesses

To early civilizations, lightning was considered so powerful and terrifying that only gods or goddesses could generate and discharge it. Their creation stories, beliefs and mythologies embodied lightning, thunder and other natural forces. Many cultures thought lightning was the visible sign of a fiery, stony weapon (a thunderbolt) thrown by the storm deity in the sky, and that thunder was the noise it created as it hurtled toward the ground. Other cultures thought lightning was discharged through the deity's use of a whip or slingshot accompanied by the distinctive crack of thunder, or thunder was the noise of the beating of drums, the turning of the wheels of a storm deity's chariot or the bellowing voice of the god or goddess. For others, lightning and thunder were the flashing eyes and beating wings of a giant Thunderbird.

Although some early cultures and civilizations extended their influence at a continental scale for many centuries and imposed their deities on those they conquered, there were many civilizations throughout the world that controlled a relatively limited area for only a century or two. People in these independent domains or city states often had their own language and developed their own mythology. Their creation myths and folklore were influenced by factors such as their physical environment and climate, the dangers they faced and the main activities that fed and sustained their communities. These early peoples experienced difficult and slow lines of communication and trade with neighbouring cultures, let alone more distant ones. With such a large number and diversity of relatively small-scale cultures and civilizations, the result was the emergence of hundreds of gods and goddesses of

*overleaf:* Early civilizations believed only a god or goddess could create lightning. Lightning near Deming, New Mexico.

9

lightning and thunder throughout the world. Only a relatively small proportion of them are well known today, while the rest have been poorly documented or forgotten entirely.

Some thunder and lightning or simply storm gods and goddesses are unique in the form they take but others resemble each other. One reason for this similarity is that it was not uncommon for conquerors to take over the deities of a previous culture while giving them their own name and weaving them into the fabric of their culture. They did this by adjusting their powers, deeds, images and worship practices to better meet their own needs. As communication and trade routes improved over time, so did the tendency for civilizations to share ideas and beliefs about storm

Assyrian thunder god, Adad, holding lightning flashes and a single-edged axe.

deities in a peaceful way, resulting in cultures incorporating and refining their original beliefs and images to take into account new knowledge and thinking. The circulation of coins helped share visual representations of the gods and goddesses, as their images were often depicted on the reverse side to the head of a ruler.[1]

An extended portfolio of responsibilities was sometimes given to the storm deities, relating to other natural elements such as rainfall and wind storms and/or to powers such as warfare (reflecting the tremendous and frightening power of thunder and lightning) and fertility (conveying the life-giving rain associated with thunderstorms). Regardless of whether or not they had an extended portfolio, storm gods and goddesses were worshipped through rituals, offerings and, in some cases, animal and human sacrifices to seek their benevolent intervention and advice or simply to appease their anger. This chapter explores the mythologies and depictions of the better documented storm gods and goddesses, the thunder weapons they employed, rituals enacted to worship them and the worshippers' beliefs about how the deities produced both thunder and lightning.

## Ancient Mesopotamia

Some of the earliest known thunder and lightning deities were powerful, masculine and worshipped by the succession of civilizations and empires that flourished in Mesopotamia, a fertile land lying between the rivers Tigris and Euphrates, now modern-day Iraq and northeast Syria. Initially, the Sumerian independent city states developed from around 3500 BC before they were united under central rule by the Akkadians between 2350 and 2150 BC. They were then succeeded by others including the Hittites, Assyrians and Babylonians. The cultures changed but there was continuity in the pantheon of gods adopted by these empires, even if their relative importance, worship practices and preferred names changed over time. The name of the storm god in the Sumerian-Akkadian-Assyrian-Babylonian succession of civilizations varied: from the Akkadian Adad (sometimes Addu), to the Sumerian Ishkur (Iškur), Teshub for the Hittites (who

assimilated him from the Hurrians), the Aramaic Hadad (Haddu) and Rammon (Rimmon), who was the Assyrian preference.[2] Their names derived from the word 'thunder' or 'thunderstorm' and the gods brought both the benefits of rain for a fruitful harvest and destruction through the effects of lightning and storms.

Marduk, the patron god of Babylon, is associated with the storm god, Adad. Here he holds a three-pronged thunder weapon in each hand.

The images of successive storm gods in Mesopotamia depicted on monuments and cylinder seals reveal developments in the depiction of lightning. Lightning was originally shown by two or three wavy or zigzag lines, representing the celestial flames of its flashes (bolts). They were later joined together at the bottom by a short stem, handle or, sometimes, a longer staff that the storm god would hold and throw. The two-pronged (bident) and three-pronged (trident) thunder weapon should not be confused with Poseidon's trident, which usually has barbs on the prongs like a fishing spear. An alternative development in imagery, along with the addition of a short stem or staff to the lightning flashes, was that two or three of the wavy lines were placed together to form a bundle of lightning flashes. The middle of this bundle was later modified and moulded together to create a handgrip and the single thunderbolt or *ceraunia* was formed, with two active ends.

The Greeks and Romans would continue the evolution of the thunderbolt by making it either simpler and smoother, like a solid dart pointed at each end, or resembling a bud or flower, sometimes with flames, wings and other ornaments added. All were throwable weapons. In some images of Adad and Rammon, the god is shown with a bundle of lightning flashes and an axe, dual weapons emphasizing the enormous power wielded by the

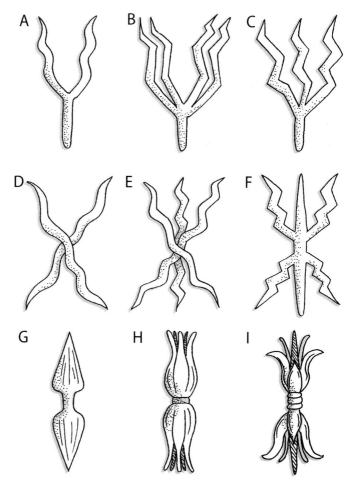

The development of lightning images.

A: lightning bident,

B: double lightning bolt bident,

C: lightning trident,

D and E: lightning flashes,

F: lightning flashes with handgrip,

G: solid thunderbolt (kerauna),

H and I: thunderbolt with a bud or flower at each end.

storm god. The single-edged or double-edged axe would, in later cultures such as in Norse mythology, come to represent the power and source of lightning and thunder alone without the addition of the older thunderbolt images.[3]

Statue of a god holding a bunch of lightning flashes outside a hydroelectric power plant, Riva del Garda, Italy.

## Greeks and Romans

From around 800 to 146 BC, ancient Greek culture flourished, with Zeus as the thunder god. He was considered the father and leader of all the Greek gods ruling Earth from Mount Olympus. In mythology, Zeus freed the Cyclopes – three one-eyed giant storm gods, who were skilled metal workers – from a dungeon. As a gesture of thanks for freeing them, the Cyclopes forged his thunderbolts, giving Zeus the power to generate thunder and lightning. Zeus is frequently depicted in one of two poses: either standing or striding forward with a thunderbolt in his raised right hand, or seated in majesty. His other symbols or attributes include a sceptre, eagle, cornucopia, ram (or bull), lion and oak tree. Lightning is often depicted in thunderbolt form both as a bulbous projectile and as a double-ended flower with several leaves or shoots above and under Zeus' grip. The Greeks recognized that Zeus used lightning not only to destroy the power of giants and monsters, but to cause destruction on Earth wherever it struck the ground: people were killed, buildings destroyed, trees blackened and toppled to the ground, and entire forests burned. Sometimes Zeus even allowed Athena, his daughter and the goddess of wisdom, to use his lightning bolts too.

Lightning features in Orphic Hymn 20:

> To Zeus Astrapaios (Lightning Maker). I call the mighty, holy, splendid, light, aerial, dreadful-sounding, fiery-bright, flaming, ethereal light, with angry voice, lighting through lucid clouds with crashing noise. Untamed, to whom resentments dire belong, pure, holy power, all-parent, great and strong: come, and benevolent these rites attend, and grant the mortal life a pleasing end.[4]

The worship of Zeus at Dodona in Epirus, in remote and rugged northwestern Greece, created the strong connection between Zeus and the oak tree, a link that would continue with Jupiter, the Roman equivalent of Zeus. Both gods were known as the Oak God and often depicted wearing a wreath of oak leaves. A sacred grove of oak trees and a shrine to the mother goddess, Dione, had existed at Epirus from as early as the second millennium BC. However, the Greeks supplanted Dione's importance at this sanctuary with Zeus and Dione subsequently became his wife. Although the foremost god at Dodona changed, similar rituals were retained. Zeus' voice, like Dione's before him, was heard in the rustling of the oak leaves, the clinking of bronze vessels hung in the branches and the cooing of doves in the grove, and these noises were interpreted as the gods' voices by the barefoot priestesses.[5] In the third century BC Dodona became so famous that it was made the religious capital of ancient Greece by the king Pyrrhus (319–272 BC), and the Temple of Zeus was rebuilt to reflect the site's huge importance. It later experienced successive acts of destruction by invading forces and was subsequently rebuilt, but by the second century AD all that remained of the sacred grove was a single oracular oak tree. This ancient tree was eventually cut down by the Roman Emperor Theodosius I (AD 379–395), who decreed that Christianity – the Catholic Church – should be the only religion practised, so he closed all pagan temples and banned all pagan religious activities.

Another important site for Zeus was the Temple of Zeus at Olympia in southern Greece, which was built in the fifth century BC and housed the famous statue of Zeus, one of the Seven Wonders of the Ancient World. The statue, showing Zeus with the figure of Nike (goddess of victory) in his right hand and seated on an elaborate throne covered by gold, ebony and ivory and inlaid with precious stones, was about 13 m (42 ft) high and attracted large numbers of worshippers. Olympia was also the site of the Olympic Games started by the Greeks and held every four years, even after Greece came under Roman rule, from 776 BC to AD 393. The games were as much a religious festival, celebrating Zeus, as they were an athletic event and consequently were stopped by Emperor Theodosius I, in his suppression of pagan worship.

Many Greek myths and gods were assimilated by the Romans, although they were modified to serve Roman needs. So Zeus eventually transformed into the supreme Roman god Jupiter. The continuing Greek influence is evident in the similar images used for Jupiter's thunder weapons. The Roman Empire, formally established in 27 BC when Gaius Octavian was proclaimed Emperor Augustus (27 BC–AD 14) by the Roman Senate, would spread the influence of Jupiter across three continents in the following five centuries. Roman emperors expected to be treated as if they were direct representatives of Jupiter and often wore crowns of oak leaves to indicate their close ties to the Oak God. Emperor Caligula (AD 37–41) even had a mechanical thunder noise-maker and spark-maker. Jupiter was believed to use lightning – his thunderbolt – as a weapon against, or omen or

Nineteenth-century wood engraving of the Temple of Zeus at Olympia.

message to, other gods and humans, and it was the responsibility of society's ruler, priests or soothsayers to interpret these.

The Romans were strongly influenced by the religious practices of the Etruscans, from Tuscany, Italy, who were at their most influential from around 600 to 480 BC, but who were eventually conquered and absorbed into the Roman Empire. The Etruscans worshipped many gods including Tinis (Tin or Tinia), their storm god equivalent to Jupiter and Zeus. One difference in the Etruscan religion was that they believed Tinis permitted nine gods or goddesses to have the power to hurl a thunderbolt. Although the Romans regarded Jupiter as the god of thunder and lightning, they sometimes attributed nocturnal incidents to Summanus, believed to be one of the nine Etruscan lightning-wielding gods. Etruscan priests had elaborate lore and practices for interpreting the meaning of their gods. They

Statue of Jupiter
holding a thunderbolt.

believed that the gods' wills were manifested through signs in the natural world and so they meticulously analysed animal entrails, lightning strikes, berries and flights of birds to discover the divine messages within. Their religious laws were codified in three sets of books. The first, *Libri haruspicini*, explained how haruspicy – the study of omens in the entrails of freshly sacrificed animals – could predict the future. The second set was the *Libri fulgurales*, which set forth the art of divination by lightning, lightning strikes and thunder, especially in terms of good and bad luck. The third set, *Libri rituales*, covered wider ritual practices as well as social and political issues.

The Etruscans believed that Tinis and, by implication, the Romans' Jupiter, employed lightning for three purposes: foretelling, frightening and destroying. Warnings were sent by Tinis alone, while lightning sent to generate fear was used on the advice of the god's counsellors and lightning sent to destroy needed the approval of the higher gods (the Fates). The crucial meaning of a lightning strike depended on the day and time, the sector of the sky from which it emanated, its colour, the duration of the stroke, its length and its trajectory, as well as whether the place where the lightning struck was a public, private or sacred building. Lightning seen in the eastern part of the sky was considered favourable while that from the northeast quadrant was especially unfortunate. The Roman historian Lucius Annaeus Seneca (4 BC–AD 65) explained that the difference between the Romans and the Etruscans was that the Romans believed lightning was caused by clouds colliding, whereas the Etruscans believed that the clouds collided in order to create lightning. In other words, Etruscans attributed everything to the gods and believed that events happened in order to express a meaning. Following their observations and interpretations of a lightning strike, the Etruscan priests would announce the appropriate responses, including the form of appeasement of the specific god or goddess who was believed to oversee that sector of the sky.[6]

The *Libri fulgurales* also instructed that the location where lightning struck should be buried or covered and a sheep sacrificed to the god to consecrate the spot. A white sheep was offered

for a daytime lightning strike and a black sheep for a night-time strike. The place would be covered by a *puteal* cylinder (lightning well) or *bidental* (more than one *puteal* cylinder, called a lightning fork) and regarded as sacred, never again to be trodden on or touched by anyone. When a building was struck by lightning, an opening was kept or made in the roof so that the god would always have ready access to the place he had chosen for himself. Throughout the Roman Empire and in many other cultures, there was a tradition that the body of someone killed by lightning should not be moved, but must instead be buried in the

Etruscans interpreted lightning according to many of its characteristics, including the sector of the sky from which it emanated.

exact spot the person had died. They were not cremated as a sign of respect for the god, whose lethal fire had killed them. Words such as 'DIVOM FVLGVR CONDITVM' (the lightning of the gods has been consigned in earth) often formed part of the inscription on the overlying *puteal* and *bidental* cylinder. Even statues had ritual burials, such as the large bronze statue of Hercules near the Theatre of Pompey in Rome, following a lightning strike.[7] Animals killed by lightning were considered sanctified by the storm god, meaning their meat was never to be eaten by mere mortals.

The *Etruscan Brontoscopic Calendar* is an almanac revealing what the rumble of thunder signifies on any given day of the year. All the daily entries begin with the phrase 'If it thunders' followed by what is expected to happen.[8] Many of the almanac entries are concerned with the likely state of harvests, such as:

> If it thunders, it signifies a good harvest (29 July)
> If it thunders, it threatens a drought. There will be an
> abundant harvest of the nut trees; around late autumn
> though, they will be destroyed by storms (24 September)

Entries also cover war, civil unrest, plagues, disease, earthquakes and other forms of disruption. For example:

> If it thunders, it threatens civil wars for the city and a plague
> on the beasts of the woods (24 December)
> If it thunders, there will be a slave revolt and recurring
> illness (7 January)

Behaviour and personal characteristics are covered too:

> If it thunders, it signals discord and thoughtlessness of men
> (18 April)
> If it thunders, men bent on vengeance shall slip into the
> worst kind of treachery (30 July)
> If it thunders, it signifies that the women are the more
> sagacious [wiser than men] (5 August)

If it thunders, the women and the servile class will dare to commit murders (19 August)
If it thunders, there shall be power among women greater than [what is] appropriate to their nature (6 September)
If it thunders, the people will be of marvellously good cheer (23 October)

The continuity of sacred sites as places of divine worship over several millennia during changing cultures and civilizations, as well as the continuity in the worship of a storm deity for a large part of that time, is highlighted in Damascus in modern-day Syria. In the ninth century BC, the Aramean king Hazael built a sumptuous temple for Hadad Rammon (the storm god is referred to as Rimmon in a reference to this temple in the King James Bible, 2 Kings 5:18). The Greeks in the Hellenistic period then used this site for Zeus. When the Romans conquered the region in the first century AD, the temple was converted to serve Jupiter. The Temple of Jupiter attracted many citizens to festivals and gatherings in Damascus, which became known as the city of Jupiter. In the fourth century, Theodosius I built an imposing church on the site, devoting it to St John the Baptist. By the seventh century, Damascus became a Muslim city and in the eighth century the site was rebuilt as the Great (Umayyad) Mosque.

## Europe

There are many thunder and lightning gods in the mythology of the northern countries of Europe, including Donar in Germany, Horagalles among the Sami (Lapps) of northern Scandinavia, Pērkons in Latvia, Perkūnas in Lithuania, Perun in Russia and Ukraine, Pikker in Estonia, Piorun in Poland, Taranis in Celtic regions such as Gaul and Ukko in Finland. Many owe their existence to the beliefs and customs that had spread from the Greeks and Romans, but these new storm gods were modified to better suit the northern people and their religions. Perhaps the most enduring was Thor (or Thunor) from Norse and Viking mythology, who by the tenth century was widely regarded as the dominant storm god throughout most of Scandinavia and its surrounding areas. Many of the other European gods share similar attributes with Thor, as they carried a short-handled axe or hammer.

Thor was depicted as a fearsome, muscular, hammer-wielding, red-bearded god. He would hurl his fiery axe or hammer, whereas Zeus or the Hindu god Indra wielded forked darts. In Norse mythology Thor's hammer is known as Mjolnir (or Mjollnir), a distinctively shaped weapon more akin to an axe than a hammer. The etymology of 'Mjolnir' is uncertain, but it is probably related to Old Norse *mala* (meaning grind), or *molva* (meaning crush), with possible cognates in the Russian *molnija* or Welsh *mellt* (both refer to lightning). Lightning is associated with the hammer, although the mythical narratives focus on the striking blow it delivers and the powers it possesses rather than the fire or light of lightning. Thunder was thought to be the noise from the rumbling wheels of Thor's chariot, pulled by goats, as it dashed across the sky, hidden from sight in the dark clouds. Thor was the son of Odin and the female giant Jörð (or Jord), symbolically considered to be heaven and earth. Thor ruled over all the features of the atmosphere, not just thunder, lightning and storms, but the life-giving rains and the winds that propelled ships across the seas. Thor used Mjolnir to rid the world of monsters and hostile giants and to protect the homes of both gods and men against the forces of darkness and chaos. Wherever Thor threw his axe it would always find its way back to his hand. It could never break or miss its target. It had immense power to crush and when striking a hill it could create deep valleys.

For many Scandinavians the protective powers of Mjolnir were considered so special that Thor's axe became a symbol used in the blessing of objects (such as ships), births, marriages, deaths and the binding of oaths. Even after Christianity replaced the old gods, it was customary in some parts of Scandinavia for the groom to carry an axe at his wedding ceremony and in Germany it was auspicious for the bride if a thunderstorm occurred during

Mårten Eskil Winge,
Thor's Battle with
the Giants, 1872.
Thor holds a
double-edged axe.

the ceremony.[9] Thor's name features in our calendar today,
with 'Thursday' being derived from 'Thor's Day'. It corresponds
to the Roman *dies Jovis* or 'Jove's Day', which was a sacred day
for Jupiter. The modern-day German version is *Donnerstag* or
'Donar's Day'.[10]

## Asia

In ancient Vedic Hinduism (1500–500 BC), the storm god Indra controlled lightning and thunder and, because of his immense physical power, he was also the god of war. Indra was commonly depicted as a muscular man with two or four long arms, golden or reddish in colour. He commanded a golden carriage drawn by two brown horses with flowing manes. In the post-Vedic period, he is shown riding a large, three-headed white elephant. In his right hand he carries the indestructible *vajra*, a highly stylized thunderbolt, although he may also be armed with a bow and arrows, a net, a huge lance and a hook to catch his enemies. The *vajra* is shaped like a double-ended flower bud or club. It may have three, five or nine spokes or prongs that usually close at each end in a lotus bud shape. The form of the *vajra* appears to have evolved from images used for the thunderbolt (lightning flashes) depicted in the ancient civilizations of Mesopotamia.

Bronze or brass versions of the *vajra* are used in Tibetan Buddhism rituals where it is called by its Tibetan name, *dorje*. The number of spokes (originally lightning flashes) and the way they come together or not at the ends have numerous symbolic meanings. The spokes of a peaceful *vajra* meet at the tip whereas those of a wrathful *vajra* are slightly splayed at the end. The *Rig Veda*, one of the oldest texts of the ancient Vedic period, identifies Indra's *vajra* as a notched metal club with a thousand prongs. Significantly, all these descriptions identify the *vajra* as having open prongs, unlike the Buddhist one, which has closed prongs. According to a Buddhist legend, Shakyamuni Buddha (also known as Gautama Buddha), who lived from 566 to 485 BC in central north India, took the *vajra* thunder weapon from Indra and forced its wrathful open prongs together, thus forming a closed, peaceful Buddhist sceptre, which absorbed the unbreakable and indestructible power of the thunderbolt. A double *dorje*, or *vishva-vajra*, is when two *dorjes* are connected to form a cross. This is associated with some tantric deities. The current national emblem of Bhutan includes a double *dorje* placed above a lotus, surmounted by a jewel and framed by two dragons, representing

Indra riding
his three-headed
elephant while holding
a thunderbolt.

the harmony between secular and religious power. Over time, Indra lost his supremacy in mythology and became subordinate to the Hindu trinity of gods known as Vishnu, Shiva and Brahma.[11]

Chinese Taoist mythology favours an entire Ministry of Thunder and Storms, comprising 24 individuals, each with some responsibilities for thunder, lightning, clouds and rain. The leader is Lei Tsu, who has three eyes, with the middle one emitting a short beam of white light. However, other Chinese thunder gods have gained greater notoriety, especially Lei Gong, also called Lei Shen. He started life as a human but ate a poisoned peach and became a fearsome creature with a blue or green body, wings, beak (or fangs) and claws who only wore a loin cloth. This god carried a drum, a hammer (or mallet) and sometimes a chisel to produce thunder and lightning to punish evildoers. According to some, it is the sound of his drums, rather than the lightning, which causes death. He dispensed justice by punishing both those guilty of secret crimes and evil spirits who used their knowledge of Taoism to harm human beings. Lei Gong received assistance from Dian Mu (Mother of Lightning), sometimes called Tien Mu, who may have been his wife. She used her ability to distinguish good from evil in potential victims and shone her twin mirrors on the intended target to ensure Lei Gong made no mistake about whom he should strike, as he had done previously without her help. Some stories claim she could generate the lightning herself by flashing her mirrors. Images of Lei Gong have changed over time: in the fourth century AD, sources depicted him as a monkey-faced man with wings and cockerel's feet, while stories from the Tang Dynasty (AD 618–907) describe him as a creature with the head of a pig.[12]

Raijin (Raiden) is the Japanese god of lightning, thunder and storms in the Shinto religion. He is typically depicted as a red demon surrounded by drums, which he beats to create thunder. Although many of his lightning strikes are unwelcome, they are believed to benefit rice crops, perhaps because they fix nitrogen, fertilizing the plants. Tradition dictates that when lightning strikes a field, farmers mark the spot with fresh-cut bamboo and rice-straw ropes to help ensure a bountiful harvest. Children are encouraged early on to fear Raijin – some Japanese parents tell their children to hide their navels during thunderstorms to avoid being taken away by Raijin, who is said to like eating the navels or abdomens of children.

Lei Gong, Chinese god of thunder, carries drums and a hammer to produce thunder and lightning to punish evildoers.

Dian Mu, Chinese goddess of lightning, holding twin mirrors to direct lightning strikes.

## The Americas

The mythology of the Inuit people who live in the Arctic and sub-Arctic areas of Canada and the United States, as well as Greenland and Siberia, attributes thunder to the goddess Kadlu. She is one of three sister goddesses who played so noisily that they were sent outside by their parents. There they invented a game to create thunderstorms. Kadlu creates thunder by jumping on hollow ice. Her sister Kweetoo creates lightning by rubbing stones against each other. The third sister, Ignirtoq, makes lightning by striking two stones together, urinating profusely to form rain. Some legends claim that Kadlu made thunder by rubbing dry sealskins together or by singing. The goddesses lived in a whalebone house in the sky, where the sisters wore no clothing and blackened their faces with soot. For food, they went hunting for caribou, striking them down with lightning. In some areas, women were said to be able to avert thunderstorms, or to create them, by leaving offerings for the weather goddesses, such as needles, bits of ivory and old pieces of sealskin.[13]

Storm deities were not always depicted in human form. Some indigenous groups in the Pacific Northwest coast, American Southwest and Great Plains thought that thunder was caused by the beating of the enormous wings of a giant, fearsome bird called the Thunderbird. It is probably based on a real bird such as a bald eagle or a hawk but became mythically exaggerated in size. Sheet lightning was created by the flashing or winking of its eyes. Individual lightning strikes were sometimes interpreted as white-hot stones shooting from the Thunderbird's eyes as fiery projectiles or even glowing snakes that the Thunderbird carried around. A legend from the west coast of Vancouver tells of the Lightning Snake who became friends with the Thunderbird and entwined itself around his body. When the Thunderbird went in search of whales for food, he would drop Lightning Snake onto the whale to pierce its body and kill it. The Thunderbird would then seize the whale in its powerful talons and return with it to its home in the high mountains.[14] When lightning struck a tree, leaving behind lines of bark peeled from its trunk, the people attributed this to the slashing of the tree by the Thunderbird's claws. Like many other thunder and lightning gods and goddesses, the Thunderbird is fierce but also a protector of humanity against evil forces. There are many stories telling of the Thunderbird's fights against malign underworld beings. In masks, it is depicted as a large multicoloured bird, sometimes with two curling horns and teeth within its long beak. On

Quillwork shoulder pouch of the Ojibwa group of Native Americans depicting two Thunderbirds.

heraldic (totem) poles of the Pacific Northwest coast it may sometimes be portrayed with an extra head on its abdomen. Individuals who survived being struck by lightning often became shamans because it was thought they had been given the Thunderbird's powers. One possible explanation of the Thunderbird belief may lie in the tendency of large birds to take advantage of the powerful updrafts that precede large storms, giving the impression that thunderstorms followed in their wake.

In Mesoamerica, during the peak of the Maya (AD 300–900) and subsequent Maya-Toltec (AD 987–1200) civilizations, the people believed that Chaac (or Chak) created thunder, lightning and rainfall in the Yucatán Pensinsula. Chaac was often depicted with a stone (serpentine) axe, which he would use to strike the clouds, triggering lightning and thunder and heralding the onset of life-giving rains after the long dry season, much to the relief of farmers. Chaac was usually shown with a human-like body, but scaled and reptilian with a inhuman head, curved

Heraldic (totem) pole with a Thunderbird at Thunderbird Park, Victoria, Canada.

fangs and a long, pendulous nose. Tears were sometimes shown coming from his goggle-like eyes to represent rain. Like other Maya gods, Chaac could also be represented as four gods, the Chaacs – one for each cardinal direction (with each having a slightly different responsibility or trait associated with each direction). Chaac was often worshipped at sacred cenotes (large, cavernous sinkholes). In the dry limestone region of the Yucatán the cenotes provided the main water source. Human sacrifice formed part of the worship of Chaac. Four priests, who were called Chaacs, like the gods, were responsible for conducting the ceremonies, which included, on occasion, boys, girls and young adults – usually male – being thrown into the cenotes to drown, along with precious offerings of gold, turquoise and jade jewellery and sculptures. At the Cenote Sagrado (the Sacred Cenote) in the major religious centre of Chichén Itzá, individuals who survived until midday after being tossed into the cenote at dawn were pulled out, as it was believed they had the power of prophecy

Temple with images of Chaac at Chichén Itzá, Mexico.

following their exchanges with those who dwelled in the murky green waters.

Chaac had a particularly strong relationship with rulers, especially during the earlier periods of Maya history, as rulers were considered rainmakers, and in later periods were able to communicate with the gods and intercede on their behalf. Much of our knowledge of the Maya gods comes from the *Popol Vuh*, one of very few Maya texts to survive the destruction of religious texts during the sixteenth-century Spanish conquest. This illustrated manuscript highlights deities who could trigger lightning, including God K (K'awil). He has the same type of zoomorphic face as Chaac, but with a mirror or a torch in his forehead and an axe blade protruding through his head. Many illustrations of God K show smoke emanating from the axe blade, which references

Chaac, with a human face, on a Maya vase holding an axe in one hand.

the fire and burning caused by a lightning strike. Sometimes one of his legs is a long serpent, which represents the lightning flash itself. God K may be shown with God N (Pauahtun), whose shout represents thunder, emerging out of the serpent's mouth.[15] While most of the ancient Mesoamerican gods are long forgotten by the descendants of the original inhabitants, prayers to Chaac are believed to continue today.[16]

The Aztecs, originating in central Mexico, came to dominate much of Mesoamerica from the fourteenth to sixteenth centuries. After two centuries of migration in search of their promised land, the Aztecs established Tenochtitlán in 1325 on a swampy island in Lake Tetzcoco. This simple settlement – today the location of Mexico City – grew into a metropolis of 250,000 inhabitants, which eventually ruled much of Mesoamerica until it was destroyed by the Spanish conquistadors in 1521. The Aztecs, or Mexica as they are increasingly being known, often adopted existing deities and integrated them into their existing pantheon. One of their most highly revered gods was Tlaloc, the god of rain and fertility, who was feared because he could command lightning and thunder. Representations of a god similar to Tlaloc and Chaac may be seen in the Teotihuacán culture of the Mexican highlands (third to eighth century AD). Such similarities point to the Teotihuacán, Maya, Toltec, Aztec and Zapotec cultures of Mesoamerica sharing a common origin for their beliefs, world views, obsession with time and calendars, and images of their gods. These links may be traced back to the Olmec civilization of Mesoamerica (around 1200 to 400 BC).

The Aztecs recognize four Tlalocs, each of whom has a distinct colour and was specifically associated with one of the four world directions. Tlaloc is usually depicted as goggle-eyed with fangs. Two serpents formed his eyebrows, or sometimes his eyes, and they intertwined to form his nose. He would often be shown wearing an elaborate headdress – with points representing the mountains where he kept the water – and large ear ornaments. The white pointy fangs have been interpreted as fangs of a jaguar, the largest predator in Mesoamerica and South America, and the rumbles of thunder were thought by the Aztecs to be its powerful roar. Some sculptures of Tlaloc were made by applying resin and copal to wood rather than clay or stone. These perishable statues were then burned after worship, once Tlaloc's favour had been gained; it was believed that the smoke created by the resin and copal would blacken the clouds and cause then to release their fertilizing rains. Perhaps this was an early attempt at cloud seeding?

The Aztecs believed that Tlaloc required the sacrifice of young children – preferably those from noble families – to ensure the dry season would end and that ample rains would then follow to provide good harvests. Children with auspicious features, such as having two cowlicks, were selected by priests at a young age and kept in a separate house until they were five or six years old. For the annual ceremony of Atl Cahualo, the month of drought, the children were dressed in colourful paper costumes, plumed with quetzal feathers and carried to one of the seven sacrificial mountain locations. People lined the routes and gave the children necklaces of green stone and the mothers and children would be observed for signs of tears. A mother's tears were considered a

Goggle-eyed Tlaloc (left) and Feathered Serpent stone heads on the Temple of the Feathered Serpent, Teotihuacán, Mexico.

Aztec Tlaloc from the 16th-century Codex Laud, folio 2. Tlaloc is holding a lightning serpent in his right hand and a stone axe in his left.

good omen and the amount of tears from the children was believed to correlate directly to the amount of rain that would fall in the season ahead. Priests would encourage these positive omens by making the children cry before being sacrificed, sometimes by tearing off their nails.

Although the Aztec dead were usually cremated, those who had been struck by lightning, as well as those who died as a result of drowning, leprosy and other water-related diseases, were buried. It was believed that Tlaloc had chosen them for an eternal and blissful life in his fertile summer paradise, Tlalocan. They were buried with seeds placed on their faces and blue paint on their foreheads. Their bodies were dressed with

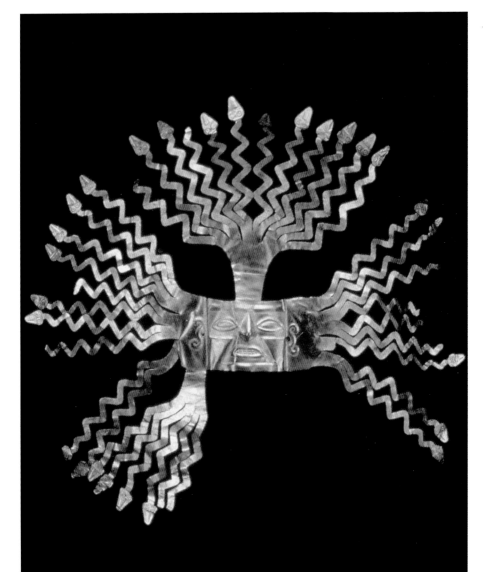

Golden mask with rays in the form of lightning serpents, symbolic of the Inca god Ilyap'a, Ecuador.

paper, and a digging stick used for sowing seeds was placed in their hands.

Before the Europeans ventured to South America in the sixteenth century AD, the Incas had established the largest-known empire, which encompassed the mountains of the Andes and the Pacific coastlands, with its capital in Cuzco. Their pantheon of gods was ruled by powerful sky deities who presided over the heavens or lived atop snow-capped mountain peaks. Inti, the sun god and divine ancestor of Inca royalty, was the most powerful but Ilyap'a (or Illapa), the god of thunder and lightning, was very important. In some South American regions, Catequil is the name used for the god of thunder and lightning, while in others Apocatequil is used when emphasizing the god's control of lightning. Ilyap'a drew water from the celestial river of the Milky Way and dispensed fertilizing rains by sending lightning to smash a great jar of celestial water. Thunder was the crack of his slingshot sending the lightning bolt, and lightning was also created by the sparkle from the movement of his brilliant clothing. Although he is pictured as a man wearing shining clothes, he was sometimes depicted as a man with a club and stones in his hands. Incas would tie up black dogs and llamas and deny them food in the hope that the cries of the animals would encourage Ilyap'a to take pity on them and send rain. The Inca made gender distinctions between 'female lightning', which strikes the ground, and 'male lightning', or cloud-to-cloud lightning. They believed that lightning affected pregnant women caught out in a thunderstorm by splitting the womb and resulting in twin births or physical deformities such as a cleft lip and palate. These children were considered special, being marked as children of lightning, and were often sacrificed later in their lives to Ilyap'a.[17]

## Africa

A storm god in the form of a bird, who can summon thunder and lightning with its wings and talons, also features in some traditional African cultures, especially in southern Africa where it is referred to as the Lightning Bird (Impundulo or Ndlati). The scars left on trees struck by lightning are thought to be lacerations caused by the talons of the Lightning Bird. It is human-sized and often depicted in variegated colours of red, green, black and white. Tradition has it that the Lightning Bird lays eggs where lightning strikes; if the eggs are not dug up and destroyed, then lightning will return to that location when the bird returns to collect the eggs.

Other African storm gods take human form. In West Africa, the Yorùbá god of thunder is Shango (or Sango), named after a warrior king of the Oyo Kingdom in the fifteenth century AD who was deified posthumously. Shango is probably the most important of the orishas, the Yorùbá deities, but he shares the power of thunderstorms with his wife, Oya, the goddess of lightning, tornadoes, winds, rainfall, fire and fertility. Shango is often depicted with an oshe, a double-headed battleaxe, which symbolizes a thunderbolt. As the god of thunder, lightning and fire, he hurls fiery stones towards the ground with Oya. Statues of Shango often show the oshe emerging directly from the top of his head, indicating that war and enemy-slaying are at the essence of his personality and fate. Oya is also the goddess of female leadership and female independence and as a fierce warrior she is called on by women in need of strength, courage and authority. When she was not in her human form she was a water buffalo; to honour her, buffalo horns are placed on her altar. In human form she is sometimes bare from the waist up, shown with a sword or machete and a turban twisted to appear like buffalo horns. She tends to be depicted cupping her breasts, sometimes looking fierce with a beard – 'the one who grows a beard to go to war'. Wherever lightning strikes, Yorùbá priests search the surrounding area for the stones thrown by the gods, looking especially for one with a double-headed axe shape, like Shango's weapon. These thunderbolts are believed to have mystical powers and were often placed in Shango's temples and shrines. During the eighteenth and nineteenth centuries, thousands of Yorùbá were taken to the Caribbean as slaves and so the worship of Shango and Oya continued in many Caribbean islands as well as some

South American countries such as Brazil, either with the same names or variations of them, such as Changó and Olla in Cuba, and Xangô and Yansā (Iansā) in Brazil.

Africa also contains one of the oldest storm gods – the ancient Egyptian god Set (or Seth), dating back to the third millennium BC, who was associated with events such as thunderstorms, sandstorms, eclipses and earthquakes. He was most often depicted as a 'Set animal' or a man with the head of a 'Set animal'. The Set animal is a dog- or jackal-like creature with a curved snout, long rectangular ears, slanted eyes, a forked or tufted tail and a canine body. In some later periods Set was depicted as a donkey or with the head of a donkey. He was also identified with hippopotamuses, crocodiles, scorpions, turtles, pigs and donkeys – all animals which were considered unclean or dangerous.

## Oceania

In Aboriginal Australian stories, lightning is a power unleashed by certain ancestral beings. The majority of aboriginal myths relate to the 'Dreamtime' or the time of creation when ancestral beings journeyed across the landscape, creating the mountains, waterholes and distinctive rock formations, generating the weather and bringing into being plants, animals and humans. Dreamtime is the basis for aboriginal culture, spirituality and intimacy with the environment. Rock drawings and paintings, tens of thousands of which have been located in Australia (some of which may date back more than 20,000 years), illustrate these ancestral beings and creation stories. Successive generations of aborigines have ensured the preservation and retouching of these important images and the retelling of the stories relating to them. Lightning ancestral beings are a common image in northern Australia, as they produce the lightning and thunder that herald life-sustaining rain. Two important lightning beings in Australia's Northern Territory were the Lightning Brothers, Yagdabula and Jabiringi, who were responsible for bringing water to the Victoria River. The striped design of their images in rock art represent falling rain. Their heads are gecko-like and their bodies are like goannas (monitor lizards) to represent them as strong, powerful creatures. Namarrgon (Namarrkun or Mamaragan), the Lightning Man, is found in Kakadu National Park, and is said to be responsible for thunderstorms in northern Australia, creating lightning by splitting dark clouds using his stone axes. His axes are often shown in paintings to be attached to his knees and elbows. A lightning strike is shown as a band stretching from his ankles to arch over his head.[18]

Maori mythology in New Zealand assumes that all natural objects and phenomena take on a human or personified form. Whaitiri is the goddess of thunder, described as a fierce, cannibalistic warrior who came down from heaven to marry a mortal but became dissatisfied with him and eventually returned to heaven. Although Whaitiri personifies thunder, each kind of thunderstorm has its own personified form, most of whom are female, although her grandson, Tāwhaki, also came to personify thunder and lightning.

In Polynesian and Hawaiian mythology, Pele (Pélé) is the goddess of volcanoes, lightning and fire. Her home is believed to be the fire pit called Halemaʻumaʻu Crater at the summit caldera of the active Kīlauea volcano. She is said to have arrived by canoe from Tahiti and dug many fire pits or craters in the islands using her digging stick. Modern images of Pele often depict the goddess in shades of red, with a digging stick in one hand and an embryonic form of her younger sister, Hiʻiaka-i-ka-poli-o-Pele, in her other hand in the form of an egg, which Pele had carried with her and kept warm in her canoe. Other images show her carrying a *palaʻā* lace fern, one of the first plants to grow on new lava. One of Pele's older brothers is Kane-hekili, the god of thunder, who expects people to show him respect by remaining silent during a thunderstorm. Pele's power to create lightning seems to be of only secondary interest in Hawaiian mythology compared with her ability to trigger volcanic eruptions – events that can have destructive and creative repercussions. Even today, some people follow the traditional beliefs by leaving offerings such as fruit, flowers and fish at the crater's rim in order to appease Pele and calm her temper but, at the same time, thank her for the new

Lightning was inexplicable to early cultures except as an act of a storm deity. Lightning in Nebraska.

land she creates when the fiery lava extends into the sea, cools and solidifies to expand their islands.

Early cultures and civilizations produced many gods and goddesses of lightning and thunder, whose forms – shown in statues, carvings, paintings, drawings and coins, or described in oral histories and documents – sometimes share many similarities, while in others they are entirely distinct. Their images may have changed over time too, even within the same culture. In past civilizations, storm deities ranked highly in the pantheon of gods and goddesses. People worshipped them in order to protect themselves against the death and destruction that lightning could bring, and because the gods' powers meant that these beings could give meaningful answers to societies' questions and satisfy their needs. Current religions relegate these storm deities to myth and folklore, but there remain some people who continue the traditional beliefs and practices, giving offerings to the storm gods and goddesses who were so important to their ancestors' lives.

## 2 Fear of Lightning: Thunderbolts, Witchcraft and Protective Charms

Many practices were undertaken by people in earlier centuries to keep them safe from lightning. Stone missiles (thunderbolts), believed to have been fired into the ground by lightning, were collected and placed in homes in the belief that they retained magical powers of protection against future lightning strikes. Witches believed to be responsible for violent weather were hunted down, tortured for confessions and killed. Church bells were rung to ward off thunderstorms and special prayers were offered to gain protection from lightning. Other practices included the wearing of charms and animal skins, growing specific plants near homes and placing cuttings or seeds of particular plants indoors. Some of these practices were encouraged by religious leaders at the time. Others originated in the folklore of one culture and were then shared and subsequently adopted by others through long-distance contact. Some of these traditional practices continue even today.

### Thunderbolts

Until the past century or two there was a widespread belief in many parts of the world that every time lightning struck the Earth, it left a stone missile or bolt embedded in or on the ground. These stones were commonly called thunderbolts (*ceraunia*) but also thunderstones, thunder axes, lightning stones and sky axes in different cultures, reflecting their association with the particular thunder weapon employed by their storm deity. They were regarded as the spent cores of a lightning discharge, thought to have magical protective properties because they retained some residual divine powers. In other words, magical power remained alive within them. So where lightning struck, people would look for traces of the strike and any unusual stone found there was often assumed to be the lightning remnant. Or people came across an unusual stone on, or in, the ground and believed it was evidence that lightning had struck that location some years previously. Indeed, there was a common belief in some cultures that it took many years, often specified to be seven, for a thunderbolt to work its way upwards to the surface following a strike.

A thunderbolt would be collected and incorporated into the walls, roofs or hearth in order to protect the house and the family against lightning as well as various kinds of misfortunes and illnesses. Hanging one in barn roofs and on animal pens would keep the family's property safe too. One was often kept on dairy shelves to prevent milk turning sour and to ensure a good cream on the milk. Its effectiveness was questionable: before refrigerators were invented milk would often turn sour or curdle during a summer thunderstorm, as the bacterium *Bacillus acidi lactici* (which produces lactic acid) multiples rapidly during a hot, humid night-time thunderstorm.

Travellers would carry a small thunderbolt in their pocket or even attach one to a cord to wear as an amulet when venturing out in a storm to ensure they were not struck by lightning. In the late eleventh century, Bishop Marbod of Rennes wrote a 732-verse book, the *Liber Lapidum* (Book of Stones), claiming:

> He who carries one will not be struck by lightning, nor will houses if the stone is there; the passengers on a ship travelling by sea or river will not be sunk by storm or struck by lightning; it gives victory in law-suits and battles, and guarantees sweet sleep and pleasant dreams.[1]

Thunderbolts were thought to have healing powers against the evil witchcraft that brought disease and misfortune. Writing in the thirteenth century, the Danish priest Henrik Harpestraeng confirmed that thunderbolts would protect against witchcraft.[2]

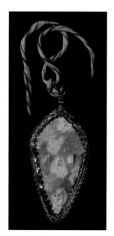

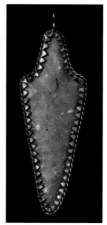

Silver mounted flint arrowheads, amulets to protect against lightning.

Parts of a thunderbolt were sometimes ground into powder and eaten as a cure for toothache, rheumatism and various other illnesses, or it was simply rubbed against the sore part of the body.[3] Animals were fed it, too, as a cure against some illnesses.[4]

One traditional belief concerning the thunderbolt's protection of a house from a lightning strike was that its presence meant the storm deity had already chosen the house, so another strike was unnecessary. Children were told to never remove the stone from the home for fear that it would be struck by lightning. Some people thought it useful at the approach of a thunderstorm to make it clear that the thunder deity had already chosen their house, so they performed a ritual imitation of the lightning flash by hurling the thunderbolt against the door.[5] The medieval Christian Church incorporated thunderbolts into their beliefs by considering them the remnants of the war in heaven, having been used to drive out Satan and his hosts. This belief explained why, in the eleventh century, an emperor of the East sent a 'heaven axe' to the emperor in the West.[6]

Generally, thunderbolts were recognized as strangely shaped stones with holes in them and/or sharp ends with polished, chipped, round and smooth surfaces. Such descriptive physical characteristics mean that a large number of thunderbolts were actually Stone Age carved artefacts, predominantly axes, hatchets, daggers, chisels, sickles, spearheads and arrowheads, that were

found but not recognized by the finders as prehistoric human-made implements. Fossils, especially belemnites (the fossilized bullet- or dart-shaped shell that formed on the inside of a squid-like creature) and echinites (the fossilized shells of sea urchins) featured as thunderbolts too, along with meteorites and crystals (for example, nodules with iron pyrites). In many cases, to be accepted as a thunderbolt an object had to resemble what people thought was the shape of the thunder weapon or missile employed by Zeus, Jupiter, Thor, Indra and other storm deities. Where a thunderbolt was made of flint it could be struck with iron or steel to produce miniature 'lightning' in the form of sparks as a clear sign of its retention of some residual lightning power. However, it was accepted that only the storm gods and goddesses could fully reactivate a thunderbolt.[7]

The thunderbolt belief was strongest in parts of the world where the knowledge of stone tools made and used by prehistoric

Natural history collections helped many people recognize that 'thunderbolts' were simply primitive stone tools. This 'Cabinet of Curiosities' was created by Ferrante Imperato (1525–1615) in Naples, 1599. Illustration from *Dell'Historia Naturale.*

people no longer existed or was simply rejected through religious beliefs, as in Europe, North America, and parts of Asia and Africa. By contrast, in Australia, South America, the South Pacific Ocean islands, and in some other parts of Asia and Africa, where knowledge of stone tools and skills was often retained, the cultures knew that stone axes and arrowheads were made by humans and so the thunderbolt myth was absent. In some of these areas, belief in thunderbolts developed when European explorers and settlers brought the idea with them from their own countries and applied it to the stone tools they found in the soil. Locations with few thunderstorms, such as Iceland, therefore had fewer thunderbolt traditions.[8]

Some theologians adopted well-reasoned arguments to reject the suggestion that thunderbolts were stone tools made by earlier primitive people. For example, in Europe in 1649, Adrianus Tollius, in his edition of *Boethius on Gems*, stated that:

> They are generated in the sky by a fulgureous exhalation (whatever that may look like) conglobed in a cloud by a circumfixed humour, and baked hard, as it were, by intense heat. The weapon, it seems, then becomes pointed by the damp mixed with it flying from the dry part, and leaving the other end denser, while the exhalations press it so hard that it breaks out through the cloud, and makes thunder and lightning.[9]

The Italian Renaissance from the fourteenth to seventeenth centuries was largely responsible in Europe for establishing the connection between thunderbolts and human-made stone tools. This period saw a rise in the study of natural history. Collections, known as 'cabinets of curiosities', were created to display and study natural specimens and cultural artefacts that had been collected from both near and remote locations and cultures. The voyages of exploration and discovery taking place at that time resulted in contact with people still using or knowledgeable about stone artefacts. Added to this, during the eighteenth century many caverns were being discovered in Europe in which human skeletons and bones of extinct animals were mingled with stone implements, which resembled the stones claimed to be thunderbolts. Consequently, thunderbolt specimens were increasingly recognized as 'human tools' of primitive peoples. However, acceptance of this claim meant contradicting the Old Testament. The findings implied the world was much older than suggested and, a theory unsupported by theologians in Europe and other Christian cultures at that time, that evolution had occurred. Challenging the Bible in this way was heresy and so many naturalists were reluctant to publish or share their views outside their immediate circle of friends. Individuals who argued openly that thunderbolts were stone artefacts of earlier primitive people were ostracized and forced to make public retractions, even fearing for their safety. Nevertheless, during the eighteenth and nineteenth centuries such an overwhelming number of claims and accompanying evidence were being put forward by respected naturalists, archaeologists and other scientists that thunderbolts were eventually recognized for what they really were: stone tools, weapons and fossils (belemnites and echinites), meteorites and crystals.[10]

Despite the growing acceptance that thunderbolts were simply primitive stone implements of an earlier age, the traditional view of thunderbolts often persisted among farmers and villagers

Most trees survive lightning strikes, but the high water content of this eucalyptus tree resulted in explosive heating and vaporization, Walcha, New South Wales, Australia.

who desired the protection they believed thunderbolts offered them. Such people could argue that a skeleton found buried with pottery shards and a stone axe meant that, rather than the person having been buried with their tools, they had been killed by a thunderbolt and buried with it. They further argued that burials of those who had not been killed by lightning may have been buried with a thunderbolt they had found during their lifetime, so that its magical properties would continue protecting that person in death from evil spirits.[11]

Given the strength of traditional beliefs and folklore it is not surprising that throughout the nineteenth century – and even into the twentieth – there were examples of people in rural communities hanging a thunderbolt under the roof of a thatched cottage or wooden barn, or carrying a piece with them during a thunderstorm, as protection from lightning. There was a tradition in parts of rural France, recorded as late as 1880, to collect and carry 'thunderstones' (*pierres de tonnerre*) in one's pockets: if thunder was heard, then the person would chant 'Pierre, Pierre, garde-moi du tonnerre!' Pierre (Peter) is also the French word for 'stone' so this chant may sound acceptable to the local clergy, as it could be considered a calling on the Christian saint for protection. In some parts of rural England, the thunderbolt was substituted with something more acceptable to Christianity: for instance, an egg laid on Ascension Day (the day on which Jesus Christ ascended to heaven, 40 days after Easter Sunday) placed in the roof of a house would protect against lightning, fire and evil spirits.[12]

With lightning producing an explosive force when striking a tree, building or soil, it was not difficult to understand why many people thought a stone missile – a thunderbolt – had hit the ground where the lightning had struck. However, any explosion associated with a lightning strike happens because moisture in a tree, wall, road or soil vaporizes almost instantaneously, due to the intense heat to which it is subjected. This causes water droplets to explosively expand as they are converted to superheated steam, which, in turn, ejects tree bark, mortar, asphalt or soil violently and noisily. There is no thunderbolt to find.

## Petrified lightning

Although there are no stone missiles to find in places where lightning has struck, on rare occasions individuals have discovered something very distinctive in the ground, which they have described as 'petrified' or 'fossilized' lightning. It occurs where lightning strikes dry sandy soils, sand pits, sand dunes and sandy beaches. There may be no obvious surface evidence but the incredible heat in a lightning discharge has created a long, narrow, glassy tube extending into the ground. Not surprisingly, those finding such a tube have described it as like holding 'spent lightning' in their hands. In exceptional cases the glassy tube may be several centimetres in diameter and many metres long, sometimes

The world's longest fulgurite, around 5 m (16 ft), being excavated by the University of Florida Lightning Research Group at the International Center for Lightning Research and Testing (ICLRT).

Lightning striking sand dunes may create hollow glassy tubes called fulgurites.

branching like the path taken by lightning in the sky. These vitreous creations are called fulgurites or, in popular terms, petrified lightning, fossilized lightning or lightning tubes. The longest fulgurite to have been excavated is around 5 m (16 ft) in length, and was found in northern Florida.[13] There appears to be no relation between the size of the tube, its bore and the thickness of its walls. Some long tubes may taper but others show no marked decrease in diameter. Not all lightning strikes on sandy soil produce a fulgurite but it seems that the more compacted and drier the soil, the greater the chances of one forming. Fulgurites are relatively fragile and are therefore difficult to extract from the soil without breaking.

Sand fused by lightning creates differing shapes and textures of fulgurites. Fulgurites are fragile and often break into segments on retrieval from the ground.

Fulgurites are created when the high temperature (1,800°C, 3,270°F) of the electrical discharge of lightning melts and fuses the silica in the sand through which it passes, creating glass. Externally the fragile tubes resemble shrivelled roots to which the sand from which it was formed has adhered, giving it a rough and irregular surface. Internally they may be smooth or lined with tiny bubbles. The glass may be translucent white, greyish or black, reflecting the composition of the sand in which it was formed. The sand on the outside surface makes it feel rough and scratchy and it may look spotted or peppered with grey or black areas. The fulgurite glass is called lechatelierite and this form of glass can be created under other conditions such as meteorite strikes, volcanic explosions and even when the current from a damaged high-voltage utility line discharges into the ground, although the latter may simply produce a surface fused linear mass or scar from the local minerals present in the soil rather than a tube extending into the ground. When bare rock is struck by lightning, a glassy surface coating together with short cracks lined with glass may form a rock fulgurite.

Fulgurites exposed by winds blowing away sand at Sibley Nature Center, Midland, Texas.

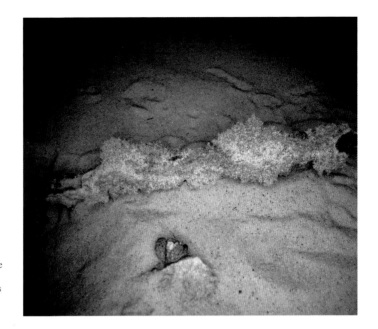

Excavation of fulgurite segments from a sand dune at the Monahans Sandhills State Park, Texas.

The earliest recorded discovery of fulgurites was believed to have been made by Pastor David Hermann in Germany in 1706, although he thought they were the product of a subterranean fire. He suggested they could be used for medicinal purposes, putting them in the same category as deer horns, crab's eyes and corals, which at that time were used to treat various fevers. The explanation that fulgurites are formed by the intense heat of lightning fusing sand grains to form glass became widely accepted when Dr K. G. Fiedler provided a comprehensive review of various discoveries of 'lightning tubes' across Europe in 1817. After that, discoveries of fulgurites increased greatly throughout the world as others searched for them in sandy soils and beaches. Their name derives from *fulgur*, the Latin name for lightning.[14]

There is no record of fulgurites being considered thunderbolts. Rather, traditional folklore in some cultures claimed they marked the passage of a fiery-hot thunderbolt that had penetrated deep down into the sandy soil – if a person dug down far enough they would be able to recover the magical thunderbolt.[15] The Native American Pueblo people in the Southwestern region of the United States, who were very reliant on rain from undependable summer thunderstorms, employed fulgurites in their rain dance ceremonies. They also carried fulgurites as charms to protect them from being struck by lightning. In many parts of the world today, fulgurites have been made into beautiful glass jewellery and are worn as lucky or protective charms.

## Witchcraft

Early Christianity in Europe sought to suppress belief in the old gods and goddesses, including the idea that gods such as Zeus, Jupiter or Thor controlled the weather. Consequently they declared that the gods of the heathen were devils and ministers of Satan. When Christian prayers for good weather for growing and harvesting crops were on some occasions answered instead by damaging drought, flood, lightning or hailstorms, it was a simple step for people to accept that weather disasters were the work of devils and demons. By the fifteenth century, this belief

A 16th-century woodcut showing lightning setting a church on fire. The lightning strikes a tree as the cattle shelter beneath it. It was believed that witches caused these misfortunes.

was extended to assume that there were individuals in their communities – men, women and children – who were collaborating with Satan to induce the bad weather and other misfortunes: these people were called witches. Fierce denunciation of witches then began, even though during the previous 1,400 years of the Christian era there had been only occasional prosecutions for witchcraft. Indeed, prior to the early fifteenth century the church had been tolerant of 'wise women' who offered folk cures or practised pagan rituals to heal the sick in local communities.

In 1437 Pope Eugene IV issued a papal bull (decree) exhorting inquisitors of heresy and witchcraft to use greater diligence against the human agents of the Prince of Darkness, and especially against those with the power to generate bad weather.[16] By 1484 Pope Innocent VIII was urging the German clergy to

> leave no means untried to detect sorcerers, and especially those who by evil weather destroy vineyards, gardens, meadows, and growing crops [and decreed that, as stated in the Bible (Exodus 22:18)] Thou shalt not suffer a witch to live.

Witch-finding inquisitors were authorized by the Pope to scour Europe, especially Germany, seeking out these evil individuals. In 1486 two of these inquisitors, Heinrich Kramer and Jacob Sprenger, published a manual in Germany – the *Malleus maleficarum* (Hammer of Witches) – which set out various means of detection and punishment of witchcraft and witches.

59

This strongly influenced Catholics and Protestants across Europe even though the Catholic Church banned it in 1490 (influence surged possibly because it had been banned). For the next two to three centuries sorcery and witchcraft were punishable by law – accusation, torture, trial and, inevitably, the death sentence – in most European countries and in the seventeenth century this extended to North America.

The torture of those accused of witchcraft in Europe forced the victims to confess and, in the eyes of their accusers, justified the drowning, hanging and burning of 40,000 to 60,000 people – around three-quarters of whom were women – at the stake, from around 1450 to 1750. Bishop Peter Binsfield (1540–1603), of Trier, Germany, was a prime mover in the prosecutions for witchcraft and he alone ordered the death of 6,000 people. He wrote a book, the *Treatise on Confessions by Evildoers and Witches*, first published in 1589, to prove that the confessions by witches under torture, including the raising of storms and controlling weather in general, were valid. Weather-related tragedies and misfortunes attributed to witches were only some of the reasons for the witch-hunts: accusations of shape-shifting, consulting with the Devil, theft, child abduction, killing of farm animals, souring of milk, damaging crops, and causing death, epidemics, illness, impotency, deformities and possession by evil spirits (bewitching) justified a far larger proportion of the deaths of those accused of witchcraft. Mass trials in Trier (1581–93), Fulda (1603–6), Würzburg (1626–31) and Bamberg (1626–31) each accounted for hundreds of deaths. Witch-hunts in Germany may have accounted for 40 per cent of all deaths of those accused of witchcraft in Europe over three centuries. At Torsåker, Sweden, in 1675 there were 71 witches beheaded and burned in one day.[17]

Witch-hunts in Scotland and England increased when James VI of Scotland (who became James I of England, Scotland and Ireland in 1603) became convinced that witches were plotting against him. This belief strengthened when a severe storm at sea endangered his life and that of his new wife, Princess Anne of Denmark, when they sailed from Denmark to Scotland in 1590. Those suspected of using witchcraft to cause this were executed

'Witches' Sabbath', woodcut by Hans Baldung (Grien), 1510. Three witches sit around an exploding potion while a fourth is seated backwards astride a flying goat.

in North Berwick after they had admitted under torture that they had tried to sink his ship. In 1597 James wrote a book, *Daemonologie*, instructing his people that they must denounce and prosecute any supporters or practitioners of witchcraft. He stated that witches

> can rayse stormes and tempestes in the aire, either upon Sea or land, though not uniuersally, but in such a particular place and prescribed boundes, as God will permitte them so to trouble (Book Two, chapter v, page 46)

Shortly after this text was published, Shakespeare reflected the views of the king, who was patron of Shakespeare's theatre company, the King's Men, by making three witches and their evil actions an integral part of his play *Macbeth*. As many as 4,000 accused witches may have been killed in Scotland and around 500 in England (including those found guilty in the Pendle witch trials in Lancashire in 1612) from the sixteenth to the early eighteenth centuries.[18]

Witch trials extended to North America in the seventeenth century with trials at Salem, Massachusetts, in 1692–3. Although there were individuals in the sixteenth and seventeenth centuries who opposed witch-hunting and challenged the validity of the confessions achieved through torture, and others who offered scientific explanations for violent weather, it was not until the late seventeenth and early eighteenth centuries that witch-hunting by Protestants and Catholics declined. Even then, some prominent religious leaders in the eighteenth century rejected the scientific explanations and continued to declare previous church orthodoxy. For example, in England in 1768 the Protestant John Wesley declared 'the giving up of witchcraft is in effect the giving up of the Bible'.[19]

Witch-hunts and the execution of witches have continued to the present day in some parts of the world. In the past 50 years, witch-hunts have led to the killing of hundreds, perhaps even thousands, of people in parts of Bolivia, Cameroon, Ghana, India, Indonesia, Kenya, Papua New Guinea, Peru, South Africa,

An accused witch directs lightning to kill her prosecutor and release her manacles, which are flung towards the judge, during the Salem witch trials, 1692–3. Joseph E. Baker, *The Witch no. 1*, 1892, lithograph.

Tanzania, Uganda, the People's Republic of Benin and Zambia.[20] According to the Tanzania Legal and Human Rights Centre (LHRC), between 2004 and 2009 more than 2,585 older women were killed in eight regions of the country because of alleged witchcraft. In traditional societies in these countries it is not uncommon for individuals, families and whole villages to believe they have suffered misfortune or tragedy because others in their community have been driven by envy, hate, jealousy or wickedness to cause them harm by their use of mystical powers and black magic. In South Africa 'almost every Zulu, Xhosa or Venda person knows that witches have the power to send lightning, kill livestock or burn houses, including the inhabitants.'[21]

Hunger, poverty, unemployment, poor education and a deep-rooted belief in traditional beliefs and customs can intensify anger at those believed to be practising harmful weather magic, as well as strengthen desire for vengeance. A traditional village healer and clairvoyant, often called a shaman or witch doctor, who is consulted by a 'victim' or their family, may encourage

63

such beliefs by confirming that someone has indeed used black magic to cast a spell on them.[22]

In Northern Province (now Limpopo), South Africa, ten villages were established by the government in 1990 as safe havens or sanctuaries to house hundreds of people accused of practising witchcraft. In one of the villages, Ebatsakatsini, which means Place of Witches, there were about 600 residents in 2010. In another village, Helena, lives a 62-year-old woman, Esther Rasesemola, whose life had been threatened in her home village. She was accused in 1990 of being a witch after lightning struck her village. Her brother-in-law, who owed her money so probably wanted to get rid of her, told the villagers and their witch doctor that she was responsible. She denied it but the villagers became convinced that she was a witch and burned down her home and took away all her belongings. Then they drove her, her husband and three children to a remote location where they left them, telling them never to return to the village or they would be killed.[23] A South African conference involving 200 police and government representatives was held in 1998 to discuss how to curb crimes by the vigilante mobs who assault, mutilate and lynch those accused of witchcraft. This conference reported that 97 women and 46 men were accused of being witches and brutally murdered by villagers or townspeople between April 1994 and February 1995. It was noted that murders are most common in the rainy season, when witches are accused of directing lightning at people that they wish to harm.[24]

According to a Congolese newspaper, on 28 October 1998, during a football game between the home team, Basanga, and the away team, Bena-Tshadi, in Kasai, the Democratic Republic of Congo, lightning killed all eleven soccer players from the away team. Another 30 spectators were injured and some were burned, but all of the home team were unharmed. This happened when the teams were drawing 1–1. Accusations of witchcraft quickly surfaced, which was of no surprise to many given that some teams in African countries have been known to use magic rituals, including the burying of animal parts and plants in and around football pitches, to weaken their opponents and give themselves a competitive edge. In a similar incident on the same weekend, lightning brought a premier league game in Johannesburg to an abrupt end after half the players from both teams – the Jomo Cosmos and the Moroka Swallows – fell to the ground. Fortunately nobody was killed but, as shown in the film of this incident, some players were left writhing on the grass holding their ears and their eyes. Several had to be carried off the field and two were taken to hospital to be treated for shock and irregular heartbeats. Some people suspected the Moroka Swallows of using witchcraft to stop the game, as they were losing 2–0.[25]

In 2003 villagers from Bokna Farm in Limpopo, South Africa, forced a 76-year-old man from his home after accusing him of using witchcraft to direct lightning on to three houses. Stones were thrown at him, he was threatened with being burned alive and the five huts he owned were torched. The previous day, a man in a nearby village of Dithabaneng was threatened with the same fate. Both men and their families sought shelter inside a police compound and the police later arrested thirteen men in connection with the Bokna Farm attack and seven people with that at Dithabaneng.[26] In 2006 two women were shot dead and a man wounded in Transkei, South Africa, in a witchcraft-related killing. This followed the death a few weeks earlier of four young men struck by lightning at a 'circumcision school'. This was the third fatal lightning strike at an initiation school in the Transkei region in three years.

Belief in witchcraft and witches and their ability to generate or direct bad weather continues today in many traditional, often rural, communities in countries across the world. Until public awareness programmes and education about the cause of lightning has displaced the traditional myths and beliefs, persecution of individuals, especially elderly women, for practising weather-magic will continue. The difficulty of overcoming traditional beliefs is exemplified by the death of Benedict Daswa. In the village of Nweli, in the Venda area of South Africa, following frequent lightning strikes in January 1990, the village headman and his counsellors were so concerned about the destruction being caused that they agreed to consult a traditional healer to

identify the witch responsible for their misfortune. For this purpose a small financial contribution per person was agreed on. Daswa, a teacher and member of the council, arrived after the decision was taken. He argued that lightning was a natural phenomenon and that witchcraft was not to blame, an opinion greeted with scepticism by the others. He refused to pay the contribution, stating that his Catholic faith stopped him from participating in anything connected to witchcraft. Many in the community believed he was belittling the traditional beliefs and conspired to get rid of him. A week later he was stopped in his car by tree logs placed across the road. When he got out of the car he was attacked by a mob who threw large stones at him. Bleeding and injured, he ran to a nearby *rondavel* (traditional round hut) for safety but the mob found him and killed him by crushing his skull with a *knobkerrie* (traditional club). Several people were subsequently arrested for the brutal murder but the court dismissed the case through lack of evidence.[27] Because Daswa died for his faith, and it was his stand against witchcraft which had brought about his death, the Catholic community proposed him for beatification and canonization. The first phase in this process was completed in 2009 and the documents submitted to the Congregation for the Causes of Saints in Rome. In January 2015 Pope Francis accepted the Congregation's recommendation that Daswa had died a martyr's death and would soon be beatified, the step before sainthood.[28]

In parts of southern Africa there are many traditional practices that village communities have adopted to protect themselves against the threat of lightning, whether it is 'natural lightning' or 'human-directed lightning', a distinction in which many community members firmly believe. A village healer (*sangoma*), who may be male or female, uses traditional medicines (*muthi*) for this purpose. This knowledge has been gained through the training they receive and the advice given to them by ancestors from the spirit world with whom they communicate, sometimes through the use of hallucinatory plant drugs. A shaman will diagnose and treat illnesses and, along with other benefits, offer ways to protect individuals against lightning and evil spirits. Traditionally, a

Traditional medicines (muthi) on sale at Mtubatuba, South Africa.

shaman protects a home from lightning by burying traditional medicines made from animal bones and plants, or placing pegs smeared with such medicines, at various points around a building. Rituals may accompany these actions.

Another role of the village shaman or 'heaven herd[er]' in Botswana, South Africa and Zimbabwe is to divert lightning away from villages. A heaven herder or sky doctor is usually male, because it is only males who herd domestic animals (or did so when they were boys) and so have developed the necessary skill. If they have previously survived a lightning strike to their home without being injured, they are believed to have been singled out for this role and are trained as a heaven herder. When a storm approaches, the shaman or a heaven herder will stand on the top of a hill near the village wearing dark clothing and pointing with a stick away from the village to redirect the approaching storm and accompanying lightning. In some cases, the pointing is done with two antelope horns filled with black *muthi* made from stones, bones, plant roots and stems. Sometimes a 'broom' of black cattle tails attached to a stick with ten black beads is used – the black is significant as it is the colour of the dark storm clouds that are being diverted.[29]

## Bell-ringing and prayers

The ringing of church bells in Christian churches has long been used to summon the faithful to religious services and, during unsettled times, to warn communities of imminent danger. However, in Europe from around the eighth century, but especially in the sixteenth and seventeenth centuries, bells were also rung to ward off the evil spirits of lightning and thus keep the church, its believers and their homes, harvests and animals safe. Bell-ringing was intended to encourage the faithful to pray even more fervently for protection, but some claimed the sound itself generated a physical force that could suppress the lightning. At the end of the seventeenth century, Father Augustine de Angelis, rector of the Clementine College at Rome, wrote:

Traditional healer whose responsibilities include herding away lightning from villages, Zimbabwe.

the surest remedy against thunder is that which our Holy Mother the Church practises, namely, the ringing of bells when a thunderbolt impends: thence follows a twofold effect, physical and moral – a physical, because the sound variously disturbs and agitates the air, and by agitation disperses the hot exhalations and dispels the thunder; but the moral effect is the more certain, because by the sound the faithful are stirred to pour forth their prayers, by which they win from God the turning away of the thunderbolt.[30]

For many centuries, the *Agnus Dei* (Lamb of God) image was impressed on consecrated oval wax discs, which were believed to protect against evil spirits, storms and lightning. Here it is carved on the wall of the Euphrasian Basilica, Poreč, Croatia.

Church bells were sometimes engraved with the words *Fulgura frango* (I break up the lightning flashes). On some bells in England, whole verses were engraved, such as:

Men's death I tell by doleful knell;
Lightning and thunder I break asunder;

On Sabbath all to church I call;
The sleepy head I rouse from bed;
The tempest's rage I do assuage;
When cometh harm, I sound the alarm.

A bell at the cathedral of Erfurt, Germany, declares that it can 'ward off lightning and malignant demons' and a bell in northeast France claims 'It is I who dissipates the thunders (*Ego sum qui dissipo tonitrua*).'[31] The prayers used to consecrate or bless the bells emphasized the importance of the protection offered by the ringing of church bells:

> whensoever this bell shall sound, it shall drive away the malign influences of the assailing spirits, the horror of their apparitions, the rush of whirlwinds, the stroke of lightning, the harm of thunder, the disasters of storms, and all the spirits of the tempest.[32]

Unfortunately the height of church towers meant they were frequently struck by lightning, resulting in the injury or death of the bell-ringers. In some instances, ironically, despite the ringing of the church bells, the tall church tower was the only structure struck by lightning during a storm. The theologian Peter Ahlwardt, in his book *Bronto-Theology, or Reasonable and Theological Considerations about Thunder and Lightning*, published in 1745, advised his readers to seek refuge from storms anywhere except in or around a church. As he reflected, lightning had only struck the churches that were ringing bells during the terrible storm in Lower Brittany on Good Friday of 1718. A study published in Munich by Johann Fischer in 1784 showed that, since 1750, lightning had struck 386 church towers, killing 103 bell-ringers. Eventually, in the face of such evidence, the Parliament of Paris issued an edict in 1786 to make the custom illegal, on account of the many deaths it caused to those pulling the ropes.[33] When lightning struck the church of St Jean on the island of Rhodes in 1856, the gunpowder stored in the vaults exploded and 4,000 people were killed.[34]

Although some religions may include prayers that praise their supreme being for thunder and lightning, some may include intercessory prayers to ask for protection against this potentially harmful and violent weather. Among several handbooks issued on this subject in the eighteenth century was one published in Zurich in 1731, the *Spiritual Thunder and Storm Booklet*, written by a Protestant scholar named Stoltzlin. It contained several hundred prayers and hymns for use during stormy weather, even including 'sighs for use when it lightens [sic] fearfully' and 'cries of anguish when the hailstorm is drawing on'.[35]

Holy water, relics and a variety of blessed objects have been advocated by religious leaders to protect against lightning, storms and bad weather. One object of the highest order for the Catholic Church, which has been in common use for over a millennium, is the *Agnus Dei* (Latin for 'Lamb of God', denoting Jesus Christ). It originated in the sixth century but became more widespread from the ninth century onward. This was a round or oval disc of wax, typically 3 to 15 cm (1 to 6 inches) in diameter, made by melting the previous year's Easter candles of Rome's churches.[36] It was blessed by the Pope's own hand, and impressed with a figure of a lamb, representing the 'Lamb of God', bearing a cross or flag. From the seventeenth century, figures of saints or the name and coat of arms of the consecrating Pope and the date were commonly impressed on the reverse. Small *Agnus Dei* discs were worn suspended round the neck, usually enclosed in a leather or silk covering, while larger ones were displayed in the home as objects of devotion. Some discs were hollow to include the gospel of St John written on fine paper. The powers of the *Agnus Dei* were considered so great throughout Europe that Pope Urban v (1363–1370) thought three of these discs a fitting gift from himself to the Greek emperor John v Palaeologus in 1366, accompanied by a letter which claimed 'It drives away thunder' as the first of its many powers.[37] In the late sixteenth century Bishop Vincenzo Bonardo of Gerace, Italy, reaffirmed that the *Agnus Dei* provided protection from lightning. Another sixteenth-century statement was that 'gainst lightning it hath soveraigne virtue, and thunder crackes beside'.[38] Only the

Pope was permitted to consecrate these wax pieces and he only performed the required ceremony in the first and seventh years of his pontificate. The traditional prayer of consecration was:

> O God ... we humbly beseech thee that thou wilt bless these waxen forms, figured with the image of an innocent lamb ... that, at the touch and sight of them, the faithful may break forth into praises, and that the crash of hailstorms, the blast of hurricanes, the violence of tempests, the fury of winds, and the malice of thunderbolts may be tempered, and evil spirits flee and tremble before the standard of thy holy cross, which is graven upon them.

The Franciscan Gabriel Sagard, in his *Grand voyage du pays des Hurons* (1632), tells that his interpreter, captured by Iroquois Indians in North America,

> was miraculously saved through the potency of the Agnus Dei that he had suspended from his neck. When the Indians wanted to tear it off, it began to thunder furiously, and it flashed like lightning, so that they thought the end of the world had come and, terrified, released him, thinking they were going to die for having tried to kill a Christian and for stealing his relic from him.[39]

The tradition of producing *Agnus Dei* discs continued until 1964, when Pope Paul VI is believed to be the last pope to have consecrated them during the Easter week.

St Barbara is a third-century Christian martyr and saint invoked by some to protect against lightning. From accounts written from the ninth century AD, it is believed she was born in modern-day Turkey. She was the only daughter of Dioscorus, who isolated her for many years in a tower to keep her away from unwelcome outside influences, including the new Christian religion. When this failed and she became a Christian, her father reported her to the Roman authorities. They shamed and tortured her to try to get her to renounce her Christian beliefs. When

**SANTA BÁRBARA** é popularmente invocada na ocorrência de grandes tempestades para afastar dilúvios, relâmpagos e trovoada. É conhecida como padroeira dos artilheiros, mineiros, bombeiros e todos quanto trabalham com explosivos. Foi colocada neste sítio para proteger os trabalhadores que aqui manusearam o material explosivo utilizado na construção deste túnel.

**SAINT BARBARA** is popularly invoked when big storms occur to keep away floods, lightning and thunder. She is known as the patron saint of the artillery men, miners, firemen and all those who work with explosives. She was placed here to protect the workers that handled explosive material used in the construction of this tunnel.

she refused, her father was ordered to kill her. After he beheaded Barbara, he was struck, burned and killed by a lightning strike on his way home. Because of her father's fate, St Barbara has since been invoked in prayers for protection from lightning as well as fire and violent, unpredictable death. Due to her association with lightning, she is the patron saint of firefighters, miners, artillery forces and many others who have dangerous professions. Her feast day is 4 December and, although the Catholic Church no longer includes St Barbara in its revised calendar list of celebrations due to the lack of evidence to substantiate her life story, she continues to be celebrated on this date by members of these professions.[40]

In 1652 the relics of St Donat (St Donatus), a second-century Roman soldier and martyr, were restored from the catacombs in Rome where they had been recently rediscovered to his birthplace, Bad Münstereifel in Germany. On the journey the pilgrims stopped temporarily at Euskirchen, about 10 miles (16 km) from their destination. The Jesuit Father Herde was sent from Bad Münstereifel to meet them and hold a mass in the town. During the mass, a severe thunderstorm developed and lightning struck the church, hitting Father Herde who, it is said, was invoking St Donat at the time. When he survived, it was deemed a miracle due to the intercession of the saint. Later that day, Father Herde led the pilgrims and relics triumphantly into Bad Münstereifel. Ever since this incident, St Donat has been invoked in prayers to protect against being struck by lightning.[41]

## Plants and animals as protective charms

St Barbara may be invoked in prayers for protection from lightning. She is also patron saint of firefighters, miners and military artillery forces who work with dangerous explosives.

It was long believed that lightning never struck certain animals and plants, and these have thus been adopted by different cultures at different times to ward off lightning and other sudden disasters, evil spirits, illnesses and misfortune. People who had to work or travel during thunderstorms would often carry or wear a cutting from a specific plant that they believed offered protection from lightning.

The oak tree has long had an association with the old thunder gods including Zeus, Jupiter and Thor, such that they may sometimes be referred to as the Oak God, as explained earlier. An oak beside Jupiter's temple in Rome on the Capitoline Hill also became revered as Jupiter's sacred tree. Many emperors and kings claiming descent from the gods have personified them by wearing crowns of oak leaves; the Druids frequently worshipped and practised their rites in oak groves. As Christianity grew in strength in Europe, Christians deliberately cut down ancient 'sacred oaks' because they had come to represent the thunder or oak gods of the older religions. Even so, the association of oak trees with thunder gods continued among some individuals and communities. This belief was strengthened by oak trees often being struck by lightning (although this is because oak trees tend to grow tall and isolated, making them more at risk of lightning strikes than other trees). There was a traditional French and Flemish belief that a piece of lightning-struck tree had magical powers and if placed in the home, perhaps even under the beds, it would ward off lightning. Perhaps this custom is one origin of the belief that lightning never strikes the same place twice.

Acorns from oak trees would sometimes be placed on windowsills to protect the household against lightning, a sign of respect for the thunder god or Oak God. If they were collected from a tree that had been struck by lightning, some people assumed their protective powers were even greater. Continuation of the acorn custom may be seen in the form of acorn-shaped knobs at the end of pull cords on curtains and blinds. Mistletoe from oak trees was believed to have particularly strong powers of protection against lightning. It was revered by the Druids, who believed it was placed there by a lightning strike from their god. In some communities, people hung mistletoe above doors and windows of a building, believing that it would protect against lightning. In Scandinavian mythology it was believed that holly also belonged to Thor – holly being particularly influential during the winter and oak during the summer. The Norse and Celts would traditionally plant a holly tree near their homes so

A *Witgat* tree is planted next to traditional village houses in the Northern Cape, South Africa, as locals believe lightning never strikes this type of tree.

that lightning would strike it, keeping their home and family safe as a result.[42] In the Khomani San community of the Northern Cape of South Africa, the community believes that the *Witgat* tree (*Boscia albitrunca* or shepherd's bush) is never struck by lightning; if caught outside in a thunderstorm, they shelter beneath it. Some have built their houses next to this tree for the protection against lightning they believe it provides.[43]

The Romans believed Jupiter was persuaded by other gods and goddesses, who considered Jupiter far too erratic in where he hurled some of his thunderbolts, to create the plant *Sempervivum hybrid* to protect them and their own worshippers from his thunderbolts. This plant is an evergreen succulent, commonly called Jupiter's eye, Jupiter's beard, Thor's beard or thunderbeard in various parts of Europe because its massive cluster of rosettes were supposed to resemble the beard of Jupiter. Its English common name is houseleek, derived from the Anglo-Saxon word *leac* meaning 'plant', such that houseleeks may have been the original 'house plant'. In North America it is commonly called hens and chicks in reference to the larger mature and smaller younger rosettes. The Romans believed the deities had instructed mortals to keep it in a vase near the door or porch of a building or even plant it on the roof for protection against lightning, fire and evil spirits. The Holy Roman Emperor Charlemagne (AD 742–814) ordered his subjects to grow these plants on their roofs to protect them from lightning. This practice subsequently spread across Europe and, through emigrants, to North America. Many towns and cities in Germany display a statue of Roland (d. 778), one of Charlemagne's military leaders. Traditionally, a houseleek has been planted in a hollow on the top of the statue's head to protect the town's inhabitants from lightning. When, in June 1907, the plant on Roland's statue outside the Märkisches Museum in Berlin died in the frost, it had to be quickly replaced.[44]

When thunderstorms threatened, Emperor Tiberius, the second Roman emperor (AD 14–37), would wear a laurel wreath – in imitation of Jupiter (and Zeus) who was often depicted wearing one – as he believed it would protect him from lightning. Laurel wreaths were highly regarded during Roman times and were an honour given to victorious individuals, including those achieving military victories. According to Pliny the Elder (AD 23–79), laurel had a 'hostility to fire', crackling upon contact, and this attribute afforded protection from lightning to wearers.[45] During the Middle Ages, laurel (or bay, which is the common name) was believed to provide protection against both lightning and witches.

In classical Greece sailors nailed sealskins or hyena skins to the masthead to ward off lightning, since it was commonly believed that lightning avoided such creatures. Some people even wore a piece of seal's skin for protection and, during a thunderstorm, it was believed a tent of sealskin provided the most effective shelter from lightning. Gaius Suetonius Tranquillus, born in AD 69, tells us in his book *The Twelve Caesars* that Emperor Augustus, the first Roman emperor from 27 BC to AD 14, was so afraid of thunder and lightning that he always took with him a sealskin cloak for protection and at any sign of a storm he would take refuge in a room which was underground and vaulted. One reason for him taking such precautions was that when he was travelling by night on a military campaign in Spain, his litter (sedan chair) was struck by lightning, killing the slave who was lighting the way. Soon afterwards he dedicated a temple in Rome to Jupiter the Thunderer. He also ordered the building of a temple to Apollo erected on a site that the haruspices (interpreters of all portents or unusual phenomena of nature, including thunder and lightning) had announced was desired by the god after it had been struck by lightning.[46]

In Roman times, a dead owl nailed to the door of a house averted all evil and misfortune its presence in the vicinity had supposedly caused. The custom of nailing an owl to a barn door to ward off evil and lightning continued into the nineteenth century in some parts of Europe. The deaths of thunder-fearing Augustus and, before him, Julius Caesar were apparently foretold by the hooting of an owl, which was widely believed to presage imminent death.[47] An old Chinese custom involved placing effigies of owls in corners of homes or on roofs to protect from fire and lightning.

## Swords, scissors and mirrors

Weapons and sharpened implements have played a protective role during thunderstorms in many cultures. Indigenous people in Canada in the early seventeenth century were asked by the Jesuit missionaries why they planted their swords in the ground point upwards, and they replied that the spirit of the thunder would keep his distance when he saw the naked blades. During thunderstorms some people, including traditional communities in Russia, would close the doors to their homes to keep demons fleeing from the thunder god from seeking shelter in their houses. They fixed scythes, edge upward, over their doors to keep out the demons. A common precaution in some European countries

Mirrors and shiny objects attract lightning, according to folklore in many parts of the world. Lightning near Socorro, New Mexico.

was to cover mirrors or turn them to face the wall, as it was believed that mirrors attracted lightning and the evil spirits associated with it. According to legend, evil spirits would hide in mirrors and when someone looked into them they would become possessed by a spirit.[48]

In European countries, in earlier centuries, it was not uncommon for people to put away scissors and knives during thunderstorms in the belief that this action would decrease the likelihood of the house being struck by lightning. In traditional Zulu areas in South Africa, there was a strong belief that lightning was attracted to white and shiny objects. White cattle were separated from the rest of the herd and white and shiny domestic implements were covered with dark blankets. To test whether such beliefs still exist today in those rural communities, a questionnaire was given to 1,050 sixteen-year-old students in the province of KwaZulu-Natal. One question asked was: 'Do mirrors attract lightning?' The overwhelming majority (91 per cent) answered yes. Another question found that 40 per cent of the respondents believed that there are people – traditional healers and witches – who can control lightning. Being aware of such beliefs is important, as many of them will need to be overcome if such communities are to adopt effective actions to reduce the damaging consequences of lightning.[49]

# 3 The Science and Nature of Lightning

There are around 800 thunderstorms discharging lightning at any one moment across the world. They produce a global lightning flash frequency of around 40 to 50 lightning events every second or around four million strikes per day, according to satellites monitoring the atmosphere.[1] Of these events, 20 to 25 per cent strike the ground and the rest are either cloud-to-air or cloud-to-cloud discharges. Lightning is an integral part of the global electrical circuit. Since electrons leak continually from the planet's surface to the air during fair weather, negative lightning discharges are needed to replenish the Earth's negative field and maintain a balanced electrical state.[2]

A thunderstorm forms from a cumulonimbus cloud several kilometres across and up to 15 km (9 miles) high. It develops and grows in height as strong air currents surge upward in the form of warm thermals, creating a tall, imposing cloud whose top is shredded into the familiar anvil shape by the high winds of the upper troposphere. A thunderstorm may produce many weather hazards including lightning, torrential rain, hail, strong winds, downbursts and, on some occasions, tornadoes. Some thunderstorms generate many thousands of lightning strikes in an hour while others produce relatively few. It is the separation of positive and negative electrical charges which leads to the creation of lightning and its audible signature, thunder.

Thunderstorms may be isolated, towering giants, sometimes developing into exceptionally strong and long-lasting supercell storms, with some of the most intense supercells forming in the American Midwest. At other times, many thunderstorms march in line abreast along the cold front of a mid-latitude cyclone (frontal depression), which sweeps forward continually replenishing the stock of warm air up-currents and cumulonimbus clouds that form the thunderstorms. Thunderstorms may also be found in tropical cyclones (hurricanes, typhoons) where they form the towering wall clouds around the eye.

Satellite-based sensors have been used to map the global distribution of lightning. They reveal that there is much more lightning in the tropical regions than the mid-latitudes and even less in the colder high latitudes. Seventy-eight per cent of all lightning occurs between 30°N and 30°S latitudes. Lightning rarely occurs in the Arctic Ocean or in Antarctica. It is more frequent over the continents than the oceans as sunshine heats up the land to higher temperatures and encourages convection and the development of more frequent and deeper cumulonimbus clouds. In many regions, lightning frequency varies diurnally and seasonally with increased lightning activity reflecting rising daily and seasonal temperatures, which encourage convection and the development of more cumulonimbus clouds.

Globally, Africa is the continent which experiences the highest frequency. Peak lightning activity is located over central Africa, where thunderstorms occur all year round as a result of moisture-laden air masses from the Atlantic Ocean rising up over mountains. The village of Kikufa in the mountains of the eastern region of the Democratic Republic of the Congo experiences the most lightning of any location on average, with 158 lightning strikes per square kilometre per year. Many local and regional factors, such as weather patterns, mountains and land-sea influences, determine why lightning may be higher or lower than the average for a specific latitude. These all play a part in influencing the frequency and intensity of thunderstorms and the accompanying lightning.[3]

The global distribution of thunderstorms is directly linked to the Earth's climate and, more specifically, the general circulation of the atmosphere. This large-scale circulation is a response to the differential heating by the sun of the tropics and higher latitudes, with the tropics providing the driving force of intense

Lightning strikes the ground in New Mexico. Only 20 to 25 per cent of all lightning hits the ground.

atmospheric convection. If Earth's climate changes, then the distribution, frequency and intensity of thunderstorms and lightning activity will change too. Climate change scenarios have been explored using powerful computer climate models and they suggest that global lightning activity will increase/decrease by up to 10 per cent for every 1°C increase/decrease in global surface temperature.

Given the current concerns about global warming, lightning activity is expected to increase in the coming decades. Because of shifts in atmospheric circulation patterns some regions may experience more lightning activity and others less but overall the total global activity will increase. The tropics, which already experience the highest frequencies of lightning strikes, are likely to experience the greatest increase in lightning activity. The models also suggest that the proportion of cloud-to-ground lightning to total lightning activity will increase. Where there are seasonal variations in the occurrence of thunderstorm and lightning activity, as in the mid-latitudes, peak activity may occur a month or two earlier and the thunderstorm season may be extended.

World map of annual lightning flash rate as observed from satellites, 1995–2003.

Lightning from a supercell storm in Nevada. Lightning flashes between parts of the storm illuminate mammatus (small globule-shaped clouds).

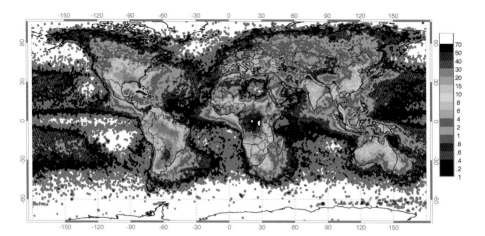

This increase in lightning activity would increase the risk of resulting damage and disruption: lightning-initiated wildfires may become more frequent, intense and extensive, especially in regions experiencing drier conditions as a result of a global warming. Although drier conditions are expected to give rise to fewer thunderstorms, computer models suggest that those which do form will be more intense. Without further investment in lightning protection systems, damage to buildings, electricity transmission lines and communication systems will likely increase, causing costly disruption to businesses and more insurance claims from businesses and homes. There is also potential for increased lightning strikes to people and animals.[4]

## Creating a lightning flash

Each brilliant flash of lightning is a transient high-voltage electrical discharge or spark whose length is measured in kilometres. Lightning ranges from 300,000 to tens of millions of volts (unit measuring the pressure with which electricity flows) with a peak current between 5,000 and 200,000 amperes (unit measuring the rate of flow of electric current) although the sustained current may be around 300 amperes. Lightning happens in milliseconds but it still generates enough energy of a magnitude to light a small city for several weeks. Its surging electrical current superheats a

Lightning heats the
air it passes through
to temperatures
hotter than the
sun. Lightning in
New Mexico.

column of air 2.5 to 5 cm (1 to 2 in.) wide to around 30,000°C (54,000°F), which is hotter than the sun. The lightning channel looks wider than it really is because it is so intensely bright. Lightning travels at about 96,000 km per second (60,000 miles per second). It produces intense bursts of x-rays and gamma rays. These invisible pulses of high-energy rays, sometimes called 'dark lightning', dissipate quickly in all directions rather than remaining in a stiletto-like lightning flash.[5]

Within a thunderstorm, positive and negative electrical charges become separated as positively charged tiny ice crystals and splinters are swept upward while large supercooled water droplets and ice pellets (graupel) carrying a negative charge descend. Charges are acquired in various ways, some of which are yet to be fully understood. Commonly it is believed that tiny ice crystals gain a positive charge when they strike large ice pellets. Also, the rapid freezing of a water droplet produces a positive outer shell and, if this fractures, tiny ice splinters carry positive charges aloft while the heavier, negatively charged inner core descends. Typically, the upper regions of a thunderstorm become predominantly positively charged (deficient of electrons) and the lower regions are negatively charged (surplus of electrons). The negatively charged thunderstorm base repels electrons on the ground away from the storm base so the ground becomes positively charged. Air between these strongly charged regions acts as an insulating layer until the electrical separation reaches about 100 million volts and lightning is triggered within the cloud, and between the cloud and the ground as the air can no longer keep the charges apart.[6]

A lightning strike to the ground begins when the negatively charged lower part of a passing thunderstorm repels electrons from the vicinity, effectively inducing a positive charge at the ground. A stream of electrons from the cloud then begins searching for a conducting route to the positively-charged surface. This faint (or visually undetected) multi-branched discharge, called a stepped leader, spreads downward in about 50-m (164-ft) spurts, carrying a negative charge towards the ground along the tips of the branching discharge and leaving a trail of ionized air.

As the stepped leader comes to within about 100 m (328 ft) of the ground, the negative charge in the air attracts short streamers of positive charges from tall objects and good electrical conductors trying to establish a flow of current. When an upward streamer connects with the stepped leader between 10 and 20 m (33 to 66 ft) above the ground, a path of least electrical resistance has been created, and a powerful bright visible channel – the return stroke – is produced.

When viewed using high-speed photography, the return stroke appears to streak upward from the ground along the channel of ionized air in an incandescent flash. This is because, once the stepped leader has created a connection to the ground, electrons in the section of the channel nearest the ground flow rapidly earthward, causing very strong luminosity. Electrons from successively higher parts of the channel then drain rapidly to the ground, such that the luminosity moves upward. Although the luminosity of the return stroke appears to move upward, the electrons in the channel are always moving earthward.[7]

The speed of the return stroke is much faster than the stepped leader. It is then followed by the first negative discharge to the surface, the dart leader. This follows the main channel in one continuous movement in contrast to the initial branching and spurt-like movement of the stepped leader. Several upward return strokes and downward strokes alternate along this channel in fractions of a second, making lightning appear to the human eye to flicker, like a strobe light. Typically there may be three or four strokes forming the lightning flash but as many as twenty have been reported.

On rare occasions when multi-stroke lightning occurs in exceptionally strong winds, the wind will blow each successive return stroke and dart leader slightly to one side of the previous stroke, producing a ribbon-like effect with the individual return strokes being separated by visible gaps. Although very strong winds may reveal the multi-stroke nature of a lightning flash, or it may even be revealed in a photograph due to the accidental movement of a fixed camera lens, it was a special camera invented by Charles Vernon Boys that enabled photographs to reveal

Positive and negative
electrical charges separate
within a thunderstorm
and eventually lightning
is initiated. Thunderstorm
at sunset in Texas.

this process so that it could be studied in detail. Boys' camera had two lenses, one stationary to capture the normal photograph of lightning, and one rotating to spread the image over a large area of the film. He invented it in 1900 but it was not until 1928 in New York State that he finally captured the first image of the multiple strokes constituting a lightning flash.[8]

## Visual forms of lightning

There are various types of visual forms of lightning. Cloud-to-ground lightning, with its intensely bright, branching channel, is what most people describe as lightning. It is commonly called forked lightning because of the many branches that are visible,

A ribbon-like lightning channel is visible when strong winds reveal the individual strokes comprising the lightning flash.

When the downward leader from the thunderstorm makes contact with an upward streamer from the ground, lightning is initiated, as here in New Mexico.

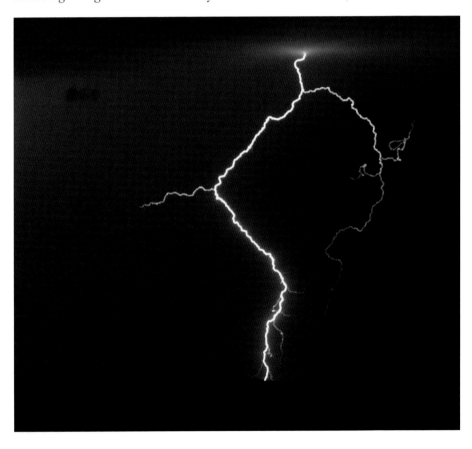

Forked lightning in Colorado, showing two ground attachment points unusually far apart.

although forked lightning may be better reserved as a term to describe cloud-to-ground lightning displaying two points of contact with the ground. This occurs when a downward stroke propagates only part of the way down the existing channel formed by the return stroke and dart leader and then forges a different path to the ground, giving rise to a lightning flash that strikes the ground in two places. It may happen because a new unattached positive streamer offers a connection as strong as the initial one during the milliseconds within which a multi-stroke lightning flash occurs. The process happens too fast for the human eye to resolve and so forked lightning is seen. The two attachment points of forked lightning are often only a few hundred metres apart.

When cloud-to-ground lightning displays one very distinct channel, with side branches being barely visible, it is some-times called streak lightning. Rare bead lightning describes cloud-to-ground lightning which appears to break up into a string of short, bright sections. This may be because the lightning channel varies in width and the wider sections fade more slowly, remaining visible for longer than the narrower sections.

Occasionally, lightning may appear to come out of a clear sky and strike the ground. This happens if the thunderstorm is hidden from view by a nearby hill and the cloud-to-ground

Forked lightning is produced when lightning displays two ground attachment points in Nebraska.

Streak lightning is
when a single channel
is very distinct with no
branches visible.

lightning travels horizontally a long distance, perhaps tens of kilometres, before descending towards the ground. It arrives unexpectedly, as a 'bolt from the blue'.

Cloud-to-cloud lightning may be within the same cloud (intra-cloud) or between different clouds (inter-cloud). Intra-cloud lightning is the most frequently occurring form of lightning and is called sheet lightning. It is seen only as a diffuse brightening of the thundercloud since the lightning, discharging from the lower parts of the cloud to the upper parts, or vice versa, is hidden within the cloud. When cloud-to-cloud or cloud-to-ground lightning is more than about 20 km (12.5 miles) away and the thunder is not audible, such brightening of the distant sky is often called summer lightning or heat lightning, so called because it occurs on hot, humid nights when thunderstorms have developed.

There are occasions when intra-cloud lightning leads to multiple lightning channels appearing to brush the irregular

Lightning travelling very long distances before striking the ground may appear unexpectedly as a 'bolt from the blue' as here in Herefordshire.

underlying cloud surface of the thunderstorm's anvil. These relatively slow-moving but visually spectacular discharges are called anvil crawlers.

Cloud-to-air lightning occurs when an electrical discharge takes place between a build-up of one type of charge within a thunderstorm and an area of opposite charge in the surrounding atmosphere. It tends not to be as powerful as cloud-to-ground lightning, and one stroke is normally enough to reduce the difference in charges to below critical levels. As a result, repeated strokes along the same cloud-to-air leader ionized path are unusual.

In cloud-to-ground lightning, short upward positive streamers connect with the downward stepped leader to initiate the lightning stroke. From very tall buildings and structures such as television towers and radio masts on mountains, a powerful upward streamer may be produced which initiates or triggers lightning from an overhead thunderstorm. This form of ground-to-cloud or upward lightning takes two forms. The more common

Multiple short lightning
channels, 'anvil crawlers',
appear to brush the
irregular underside
of the anvil cloud in
Denton, Texas.

form consists of a single, branchless leader rocketing upward from the structure tip, while the more visually spectacular but less common form consists of a branching network stretching upward from the tip of the structure – the inverse branching image of cloud-to-ground lightning. As the lightning discharge continues, the number of branches diminishes until only one or two main channels remain to carry secondary return strokes. These forms of upward lightning happen only because of the intense streamers flowing from the tip of the tall structure. Depending on the frequency of thunderstorms in their vicinity, very tall structures may be struck by lightning 50 to 150 times a year, the majority of which are initiated by the structure as ground-to-cloud lightning rather than cloud-to-ground lightning.[9]

The globule-shaped underside (mammatus) of an anvil cloud in Oklahoma, is shown lit by this intra-cloud lightning. Lightning contained within the thunderstorm results in no channel being visible from the ground.

Cloud-to-air lightning peters out in the surrounding sky.

Upward lightning triggered by the 312-m- (1,024-ft-) high Suchá Hora television and radio transmitter on the Kremnica Mountains, Slovakia.

In rare cases a positively charged giant 'superbolt' may discharge from the forward (or rear) anvil cloud to the negatively charged ground ahead of (or behind) the thunderstorm base. Positive lightning discharges (positive polarity) represent only one in twenty of all lightning strikes. The superbolt may be particularly dangerous because the distance between the anvil and the ground means it is triggered only when very large voltage differentials are overcome and so it carries six to ten times

Upward lightning from the 324-m- (1,063-ft-) high Eiffel Tower, Paris, in 1902 – one of the earliest photographs of lightning in an urban setting.

Lightning from the anvil cloud of a thunderstorm is usually very powerful (a 'superbolt'), as here in Oklahoma.

the electrical charge of the more common negative lightning discharge. A superbolt from the rear anvil is additionally danger-ous because people may think the lightning activity from the thunderstorm has passed but, instead, a superbolt brings an unwelcome 'sting in the tail'.

A superbolt from the thunderstorm rear (or forward) anvil cloud may strike the ground at a considerable distance from the core of the storm, surprising many people who think the storm had passed (or not reached them yet). Lightning in Nebraska.

## The sound of thunder

The sound of thunder arises because the superheating of the air through which the lightning travels causes the air to expand explosively and then contract rapidly, producing sound waves. Lightning travels many kilometres through the air and the rumbling sound or peal of thunder is caused by hearing the sound of thunder from segments of the branching lightning channel at varying distances from your location. The orientation of each segment and acoustic echoing from nearby hills are other influences on the nature and duration of the sound. The duration of thunder is also due to the varying time taken by sound to travel through layers of atmosphere of different temperatures. A low-intensity hissing and sharp crack or click usually indicates that lightning is relatively close to you. If further than 1 km (half a mile) away, thunder typically consists of a rumbling noise punctuated by several loud claps. A loud explosive boom, felt as much as it is heard, is often due to lightning at high altitude.

Light travels a million times faster than sound so lightning is seen before thunder is heard. The distance to lightning can be estimated by counting the number of seconds between seeing lightning and hearing thunder (the flash-to-bang method) and dividing the number of seconds by three (five) to get the distance in kilometres (miles). For example, if you see lightning and it takes fifteen seconds before you hear the thunder, then the lightning is 5 km (3 miles) away from you. Beyond about 15 to 20 km (9 to 12.5 miles), thunder may not be audible. The explosively expanding air creating the noise of thunder from nearby lightning – an acoustic shock wave – may shake cars and trigger their alarms, rattle and even shatter windows and cause ornaments on shelves and pictures on walls to be dislodged.[10] Yet, as Mark Twain wrote in 1908: 'Thunder is good, thunder is impressive; but it is lightning that does the work.'[11]

## Lightning as a form of electricity

Lightning is a giant electrical spark generated by processes occurring in a thunderstorm. A few ancient philosophers had attempted early scientific explanations, such as Aristotle around 350 BC who suggested that a 'thunderbolt' (lightning) is created when a hot, dry exhalation is squeezed from a cloud. Pliny the Elder (AD 33–79) suggested that 'by the dashing of the two clouds, the lightning may flash out, as is the case when two stones are struck against each other.'[12] However, it was not until the eighteenth century that scientists demonstrated correctly that lightning was an electrical discharge. For thousands of years, most people had simply accepted that lightning was created by the magical powers of their storm god and goddess or, as modern religions replaced these ancient ones, was the result of either divine providence or diabolical agency – a kind of meteorological struggle between good and evil.[13]

The foundation for scientific explanations of lightning came when William Gilbert, physician to Elizabeth I, showed that glass, amber and other substances attracted feathers and other objects and materials when rubbed with silk, wool or leather. Although this effect had first been demonstrated by the Greek philosopher Thales around 600 BC, Gilbert is most associated with it because he called this attractive force *electric*, from the Greek word for amber, *electron*. By the mid-eighteenth century, frictional electrical machines had been invented that could generate static electricity: a cylinder or sphere of glass or sulphur was mounted on an axle that was rotated while rubbing against pads of leather. This prompted demonstrations of 'electrical magic' in which bells were made to ring and various properties of electricity were explored. On 14 March 1746, at the Château de Versailles, Abbot Nollet entertained Louis XIV and his courtiers by arranging for up to 140 people to join hands in the Hall of Mirrors to share the shocking experience of what was at the time called electrical 'concussion'.[14]

Benjamin Franklin (1706–1790) was the person who explained, and provided the scientific experiment to prove, that

Benjamin Franklin conducting his experiment with a kite in Philadelphia in June 1752 to prove that thunder clouds were electrified and that lightning was 'electric fluid'.

soldier to stand in a sentry box as storm clouds developed. A 13-m-long iron rod resting on a low wooden platform extended above the sentry box. A glass bottle insulated it, preventing the flow of electrical current, at the base. With the thunderstorm developing, the soldier held a wire that was connected to the ground at one end, so that an electrical charge would dissipate into the ground, and touched the other end to the rod. Sparks were created and the soldier, perhaps somewhat surprised, survived.

Franklin's studies of lightning and electricity helped him realize that lightning would follow the path of least resistance as it approached the Earth, seeking out materials and objects that were good conductors of electricity. He explored ways of drawing the lightning from a thunderstorm to conduct it safely to the ground without causing any harm. In September 1752 he fitted an iron rod with a sharp pointed tip to the chimney of his house in Philadelphia. The metal rod extended about 3 m (10 ft) above the chimney. The rod had an insulated wire connected to it and passed down the stairs inside his house to an iron water pump, which would ground the electric current. On the staircase, he divided the wire by about 12 cm (5 in.) and placed a bell at each end (one of which was grounded), and between the bells a small brass ball was suspended by a silky thread. When electrified clouds passed overhead, the ball moved between the bells, striking them, and indicating that his lightning rod or conductor was safely channelling electric current to the ground. Franklin commented:

> I found the Bells rang sometimes when there was no Lightning or Thunder, but only a dark Cloud over the Rod; that sometimes after a Flash of Lightning they would suddenly stop; and at other times, when they had not rang before, they would, after a Flash, suddenly begin to ring; that the Electricity was sometimes very faint, so that when a small Spark was obtained, another could not be got for sometime after; at other times the Sparks would follow extremely quick, and once I had a continual Stream from Bell to Bell, the size of a Crow-Quill: even during the same gust there was considerable variation.[16]

Franklin's wife, Deborah, was so disturbed by the ringing bells and flashes of light that she wrote to her husband when he was in London, asking him how to disconnect the experiment.

Benjamin Franklin announced his invention of the lightning rod, which would protect homes and ships from lightning, in the 1753 issue of his *Poor Richard's Almanack*:

How to secure houses, etc. from Lightning

It has pleased God in his goodness to mankind, at length to discover to them the means of securing their habitations and other buildings from mischief by thunder and lightning. The method is this: Provide a small iron rod (it may be made of the rod-iron used by the nailers) but of such a length, that one end being three or four feet in the moist ground, the other may be six or eight feet above the highest part of the building. To the upper end of the rod fasten about a foot of brass wire, the size of a common knitting-needle, sharpened to a fine point; the rod may be secured to the house by a few small staples. If the house or barn be long, there may be a rod and point at each end, and a middling wire along the ridge from one to the other. A house thus furnished will not be damaged by lightning, it being attracted by the points, and passing thro the metal into the ground without hurting any thing. Vessels also, having a sharp pointed rod fix'd on the top of their masts, with a wire from the foot of the rod reaching down, round one of the shrouds, to the water, will not be hurt by lightning.

By 1762 evidence of the performance of various lightning rods installed on an increasing number of buildings had helped Franklin improve the recommended design and these details largely remain the basis for all modern lightning protection codes in the world today. In 1776 Franklin's legacy in science and international diplomacy was captured in an aphorism: 'He snatched lightning from the sky and the sceptre from tyrants.'[17]

Not everyone welcomed Franklin's lightning rod. Religious leaders initially objected to Franklin's attempts to 'control the artillery of heaven' and this delayed for many years the installation of lightning rods on church towers and other buildings. Those who did install them on their homes faced threats to have them forcibly removed because of religious objections by their neighbours. The religious resistance to rods focused on the belief that thunder and lightning were the will of God, often

Benjamin West, *Benjamin Franklin Drawing Electricity from the Sky, c. 1816.*

Lightning in New Mexico.

signifying his displeasure, so should not be prevented even if this meant damage to a building. Over time, acceptance of the benefits of the lightning rod increased throughout Europe and America as evidence accumulated of lightning being safely conducted to the ground without damage to the building or its occupants. In particular, tall church towers no longer had to be repaired frequently as a result of damage by lightning strikes and those attending services no longer feared for their lives at such times.[18]

Lightning formed during an eruption of Eyjafjallajökull volcano, Iceland, 17 April 2010.

## Thundersnow and volcanic lightning

A snowstorm may produce lightning and thunder, an event called thundersnow. This occurs only on rare occasions, perhaps in less than one in a hundred snowstorms. The winter surface air associated with a snowstorm is usually too cold to generate the strong upward motions needed to produce a thunderstorm within which swirling ice crystals, water droplets and ice pellets create the electrical charge separation leading to lightning. However, there are occasions when a source of relatively warm, moist air enters a snowstorm and triggers the strong convection needed to initiate the thunderstorm electrification processes: for instance, when a winter storm passes over the relatively warm waters of large lakes, such as the Great Lakes of North America.

Lightning can occur through natural means other than the usual thunderstorm mechanism for separating positive and negative electrical charges in the atmosphere. Spectacular lightning may occur in the turbulent debris-laden atmosphere above a large volcanic eruption, whether or not cumulonimbus clouds are formed. One explanation is that it may be initiated when different sized and shaped ash particles become electrically charged through friction during the eruption and then are separated: larger positively charged particles move downward and smaller negatively charged particles upward in the ash cloud.

Wildfire smoke plumes may develop into pyrocumulonimbus, producing lightning but little or no rainfall, as here at Kaibab National Forest, Arizona, in 2006.

# 4 Lightning Threats to People and Activities

Lightning poses a significant threat of injuries and death to people throughout much of the world and the economic costs of lightning damage, disruption and protection measures for many countries are enormous. Vast areas of forest and grassland and many hundreds of homes may be destroyed by wildfires triggered by lightning. Buildings and indoor electrical equipment may be damaged by electrical surges and fires. Businesses and homes may lose their electricity supply through damage to the power grid infrastructure. Ships, aircraft and spacecraft may be damaged. The health costs of dealing with lightning victims, the disruption and delays caused by lightning to national and international economies, and the huge costs of investment in lightning protection measures mean that lightning continues to be a major weather hazard.

## Injuries and loss of life

Many tens of thousands of people may be injured and killed by lightning each year throughout the world. Based on national data from 2000, tropical and sub-tropical African, South American and Asian countries feature among the worst for lightning-related deaths, including Brazil with 0.8 deaths per million of national population (based on the period from 2000 to 2009), Cambodia (7.8 for 2007–11), Colombia (1.8 for 2000–2009), Malaysia (0.8 for 2008–11), Swaziland (15.5 for 2000–2007) and Zimbabwe (14 to 21 for 2004–13). In contrast, more industrial European nations and countries such as Australia, Canada,

Japan and the U.S. currently experience death rates of less than 0.5 deaths per million or even 0.1 deaths per million in some cases.[1]

These national death rates from lightning reflect the frequency of lightning strikes, the number of people working in outdoor occupations – especially labour-intensive agriculture – the strictness with which health and safety protective measures are applied to outdoor occupations and buildings, the availability of prompt and effective medical treatment for lightning casualties, and the success of educational and public campaigns in encouraging people not to put themselves at risk of being struck by lightning.

Most lightning-related deaths have decreased over time as countries have become more industrialized, employing more people in factories and offices in urban areas rather than in exposed outdoor rural occupations. Working and living in urban areas means that most people are protected by substantial buildings in contrast to traditional wooden and thatched dwellings, which offer little or no protection to the occupants from lightning strikes. In October 2014 lightning killed eleven people and injured fifteen more when it struck a thatch-roofed, adobe-walled ceremonial hut near the town of Guachaca in Colombia's Sierra Nevada Mountains. They were among 60 people gathered in the hut for a religious service. The United States and European countries have experienced marked reductions in lightning deaths in the past century. For example, the average rate of lightning deaths per million people per year in the U.S. fell from 4.8 per million in 1900–1909 to 0.2 per million in 2000–2009. In actual numbers that has meant a reduction from over 450 people in some years of the first half of the twentieth century to fewer than 30 in the most recent years. In the UK the rate fell from 0.39 per million in 1900–1909 to only 0.02 per million in 2000–2009. Actual numbers in some years of the early nineteenth century in the UK were around 30 deaths per year but, in the past decade, there have been some years in which nobody has died. The number of people injured by lightning may be ten to twenty times the number killed.[2]

National death rates from
lightning have fallen
as populations have
moved from rural
to urban areas.

## Lightning effects on people

Although lightning is a potentially lethal high-voltage electric current, it has contact with a person for only a fraction of a second, often not long enough to cause the skin's resistance to break down and allow the current to enter the body. The fleeting nature of lightning contrasts with the longer-lasting effects of the continuous current experienced if someone touches a bare wire of an electrical circuit in a building or an outdoor power cable. The briefness of the lightning discharge may result in the

Lightning in New Mexico.

majority of the electrical current 'flashing over' the outside of a person's skin or clothes, especially if they are wet, rather than all or most of the current penetrating into the skin, passing through the body and causing serious injury by stopping the heart or damaging the lungs, brain or other organs. This does not mean that wearing a wet raincoat provides protection against lightning. Notwithstanding this myth, the surface 'flashover effect' explains why so many people survive a lightning strike.

To demonstrate how wet objects conduct electric currents far more effectively than dry objects, a laboratory experiment at Culham, near Oxford, generated and applied 200,000 amperes to a freshly cut tree stump. When it was dry, the wooden stump disintegrated explosively, but when wet – claimed to be analogous to a rain-soaked person – it suffered only minor external tearing of the bark because the surface flashover effect took place.[3] Benjamin Franklin was one of the first to note that wet clothes provide a conducting pathway outside the body if lightning strikes. His laboratory analogy was that 'A wet rat can not be kill'd by the exploding electrical bottle, when a dry rat may.' Or expressed in today's language: 'You cannot electrocute a wet rat.'[4]

Some people struck directly by lightning may suffer severe burns but many others emerge relatively unscathed, having experienced only a small electrical shock and minor burns. Why a direct strike leads to some or all of the electrical discharge penetrating the body rather than flashing over it is not fully understood. Similarly documented lightning incidents of almost identical situations may have very different outcomes for the individuals affected. Some of this may be explained by the magnitude of the voltage and current, and the differences and duration of individual cloud-to-ground lightning discharges, since there is much variation, but there is clearly much yet to learn.

A broad band of superficial or partial-thickness burns on a person arises when the heat generated by lightning passing over the skin causes the moisture in sweaty parts to reach boiling point and vaporize. The electrical current may follow lines of sweat as it makes its way to the ground, as a moist skin surface is a better conductor than surrounding dry skin. Burns may be evident in lines, typically up to about 4 cm (2 in.) wide, or areas of sweat such as down the mid-chest or spine, beneath the armpits, beneath the breasts and around the groin. Linear burns may continue down one or both legs as the current seeks to earth into the ground.

If the sudden explosive vaporization of moisture on the skin takes place on sweaty feet then, in the confined space of tight-fitting and relatively dry footwear, socks may be ripped and the footwear torn apart. This happened to a sixteen-year-old Swedish girl playing football at Gävle, near Stockholm, in July 1994. The match had only been in progress for a few minutes when a thunderstorm moved into the area. She was struck by lightning, which caused cardiopulmonary arrest for four minutes before she was resuscitated. She had been blown out of her boots and hurled into the air. Her boots and shin-guards were torn to shreds and set alight. She suffered serious burns over her body and deafness. Despite this terrifying ordeal, when she regained full consciousness in hospital her first question was: 'Who won?'[5]

Lichtenberg figure on the arm of a 10-year-old girl who was touching a metal sink when lightning struck her house in Glasgow, Scotland.

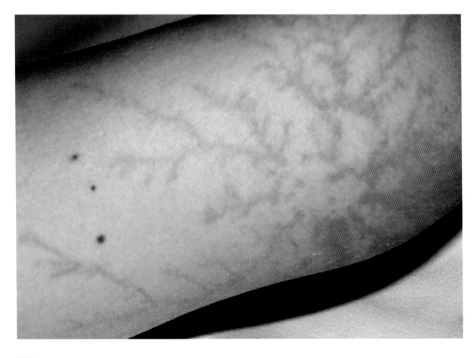

Woman with
Lichtenberg figures
on her neck, abdomen
and arms after lightning
struck her house while
she was using a corded
telephone in Galway,
Ireland.

People who suffer serious, full-thickness burns (which destroy both dermis and epidermis) or lose consciousness, without cardiopulmonary arrest, are highly unlikely to die, although they may have serious and long-term injuries. When immediate death occurs it is not caused by burns, as many people might think, but cardiac and respiratory arrest. If lightning causes cardiac arrest, a healthy, well-oxygenated heart's automaticity will usually cause a normal heartbeat to resume spontaneously within a short time. However, respiratory arrest caused by paralysis of the medullary respiratory centre in the brainstem may last far longer than cardiac arrest. Consequently, unless the victim receives immediate ventilatory assistance, attendant hypoxia may induce arrhythmias and secondary cardiac arrest. This highlights the importance of initiating cardiopulmonary resuscitation on a victim, although generally a pulse needs to be regained within 20 to 30 minutes if they are to survive.[6]

On some occasions, unique evidence of the flashover effect is evident in the form of superficial pink or red fern-like branching patterns on a person's skin. Georg Christoph Lichtenberg (1742–1799) first described this startling phenomenon in 1777 and the term Lichtenberg figure has been used to describe this pattern since 1976. It is not a burn and occurs only in people struck by lightning, although the pattern can be artificially created. The figures may occur on the torso and limbs but usually spare the face, hands and feet. It may not be visible until half an hour or even several hours after the lightning incident and it fades with time, lasting a few hours to 24 hours or even 48 hours. In exceptional cases, it may still be visible after several days. It is described as a fractal pattern, which is one of repeated bifurcations, rather like a dendritic drainage pattern seen from the happened if they had experienced a direct strike). However, in some cases an individual was the lightning strike point, so they may suffer serious injury or even death from the direct strike, while many others nearby are thrown to the ground and often experience only minor injuries.

On occasion, when lightning strikes the ground, instead of continuing its penetration of the ground and generating a ground current, it may arc across the ground surface in a ray-like or sunburst pattern from the strike point. The radial surface arcing pattern is sometimes made visible afterwards where grass has been scorched at distances of 5 to 10 m (16 to 33 ft). Standing on a surface arc of concentrated electrical charge may give rise to larger shocks to people than are experienced from a more dispersed outward-moving ground current.[18]

When a thunderstorm passes over an area, it attracts upward streamers of electrical charges from objects on the ground. One of these streamers may eventually attach to the downward leader to initiate lightning but many others may fail to make contact. These unconnected upward discharges may cause hair to stand on end and objects to buzz or crackle. In extreme cases the electrical charges being generated may be strong enough to cause minor injuries. Unconnected streamers are passing upward into the air from the highest part of the body, so the electrical effects of intense streamers on the eyes and brain are of concern. Fortunately the electrical current from unconnected streamers is ten to 50 times less than that generated by, say, a direct strike.[19]

A mountaineer climbing one of the tallest peaks in Ecuador in December 2011 recalled near the summit:

> The metal on my trekking poles – which were on my rucksack – started to buzz. Then the carabiners (metal clips) on my harness started buzzing too. The air was charged with electricity and a lightning storm was imminent. We started a quick descent back to the mountain refuge. Everything was buzzing and I felt a hot burn on the base of my back where a carabineer sat. It felt as if I was touching an electric fence, so descended even faster before being fried.[20]

At Sequoia National Park, California, the hair of siblings Sean, Mary and Mike McQuilken stood on end just before lightning struck them. Sean was knocked unconscious but resuscitated by Mike. Another person nearby died.

## Struck more than once?

In 2013 a troubled Colombian man, Alexander Mandon, who was struck by lightning four times in only six months, took the advice of a traditional healer to try to avoid being struck again. He was buried up to his neck for several hours so the soil would 'absorb inappropriate electrical charges in his body'. He was subjected to the ordeal twice because villagers first buried him lying down but were then told it had to be upright for the 'cure' to work.[24]

Melvin Roberts of South Carolina has survived being struck six times (so far). In June 2011 his latest strike left him with burns on his legs and feet and a scorch mark where he wore his watch. His previous strike in 2007 left him in a wheelchair for a year. It seems when a storm is approaching his overriding concern is to go outside to cover up his outdoor equipment and chickens from the rain rather than consider the lightning risk to himself.[25] Carl Mize, from Oklahoma City, was struck by lightning six

Screen shot captures lightning strikes in the southern hemisphere in summer when thunderstorms are active. Global, national and local lightning detection systems are now widely available using ground-based sensor networks and satellite sensors.

times between 1978 and 2006. He has suffered burns and, after one strike, was kept in hospital for four days. Since his last strike he now seeks shelter when a storm is approaching.[26] A park ranger in Virginia, Roy Sullivan, was struck seven times during his lifetime. He was first struck in 1942 and lost the nail of his big toe, then again in 1969 when his eyebrows were burned, in 1970 which left him with a seared left shoulder, in 1972 when his hair was set on fire, in 1973 when his re-grown hair was set alight, in 1976 with an injured ankle, and in 1977 with chest and stomach burns. In 1983, at 71 years old, he died from a self-inflicted gunshot wound.[27]

## Living with lightning and reducing personal risk

Smartphone apps are available which plot lightning strikes in your vicinity and, if they pose a threat, will alert you to 'seek shelter now' as shown here for Boynton Beach, Florida.

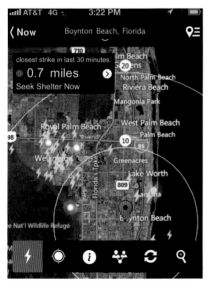

Lightning is dangerous and we need to learn to live with it. We need to ensure that nobody exposes themselves unnecessarily to lightning when thunderstorms are forecast or are developing in the vicinity. Thunderstorm forecasts need to be considered when planning outdoor work and leisure activities. The key to lightning safety is simply avoiding being in the wrong place at the wrong time.[28] Fortunately local and national forecasts and warnings have improved considerably in recent years and global and national ground lightning detection networks and satellite-based sensors can provide continual updates of lightning locations. This information can be communicated to people and organizations through television, computers, tablets, radio and mobile phones. Applications for smartphones may show the locations of lightning strikes in the vicinity of the user and may advise the user to 'seek shelter now' if lightning is too close. Other forms of hand-held early warning lightning detectors are available too.[29]

To help decide when lightning poses a serious threat, the 30–30 rule has been widely adopted. People are advised to seek shelter if thunder is heard within 30 seconds of seeing the lightning flash. A time of 30 seconds or less from 'flash-to-bang' indicates that lightning is dangerously close – within 10 km (6 miles). Although there is a small risk of lightning striking in the vicinity even when the 'flash-to-bang' exceeds 30 seconds, this timeframe is considered sufficient to cover the greatest risk. The other part of the 30–30 rule is to wait a full 30 minutes from the last lightning flash before resuming outdoor activities ('Half an Hour since Thunder Roars, Now It's Safe to Go Outdoors!'). Too often, lightning victims failed to realize the long distances lightning may travel before striking the ground, and mistakenly thought that because the heavy precipitation and dark thunderstorm base had moved away from the immediate vicinity, they were safe.

Public educational information about the threat posed by lightning is widely available on the Internet. Intensive public campaigns may be conducted on anniversaries of particular lightning tragedies, following the first lightning casualties of the year or at the start of the thunderstorm season if there is a marked

Successive strikes may occur some distance away: the 30–30 rule is to remain in shelter at least 30 minutes after the last lightning flash. Lightning in New Mexico.

One of many lightning posters encouraging lightning safety awareness.

seasonal distribution. Advice and guidance are targeted especially at those outdoor occupations and leisure activities that place people at the greatest risk from lightning. The u.s. holds a National Lightning Safety Awareness Week in June each year, with each day of the week focusing on a different aspect of the lightning threat and ways to lessen the risks posed to individuals and groups. Simple but memorable guidance is provided in the form of brief rhymes such as 'If you can see it [lightning], flee it; if you can hear it [thunder], clear it', 'Don't be lame, end the game' and 'Don't be a fool, get out of the pool.' Sports clubs are advised to develop a lightning safety policy and plan to minimize risk and incidence of injury.[30] Curtailing or delaying an outdoor activity when lightning threatens could save lives.

In some countries, lightning detection and warning systems have been installed at golf courses, parks, schools, outdoor public swimming pools and other public facilities. These may be linked to a regional or national detection and forecast centre or may be free-standing systems. Florida leads the way, as it experiences more lightning strikes, deaths and injuries than any other u.s. state. Typically a fifteen-second-long warning from a horn or klaxon is sounded a few minutes before lightning may be expected in the immediate vicinity. Those who hear the warning are expected to cease their outdoor activity and seek shelter in a nearby substantial building or enclosed motor vehicle. A flashing light attached to the klaxon may continue flashing during the time of the lightning threat and, when it is safe, three five-second warnings are sounded to indicate that the risk of being struck by lightning has passed.[31]

Golf courses in many countries have installed warning systems which, if activated, require their members to discontinue play and return to the clubhouse, avoiding tall trees and metal fences en route. Unfortunately some golfers have failed to heed the warnings and advice and, with unfortunate consequences, continued their round or taken shelter under a nearby tree. Although golfers have featured prominently as victims of lightning in the u.s. in the past three or four decades, the situation is changing. From 2006 through 2012, 238 people were struck and killed by lightning in the u.s., with almost two-thirds occurring among people who had been enjoying outdoor leisure activities at the time. During this seven-year period, people fishing accounted for more than three times as many fatalities as golfers, while camping and boating each accounted for almost twice as many deaths as golf. From 2006 to 2012, there was a total of 26 fishing deaths, fifteen camping deaths, fourteen boating deaths and eleven beach deaths. Of the sports activities, American football saw the greatest number of deaths (twelve), as compared to golf (eight).[32]

## Places to avoid – or hide – during thunderstorms

Locations to avoid when thunderstorms develop include mountains, hilltops and moors or wide, open spaces such as agricultural fields, sports fields, golf courses, beaches and open water, because a person may be the tallest point in the vicinity. Open fields give rise to the most deaths in Bangladesh as farmers and farm labourers make desperate attempts to harvest the rice crop and save it from flooding and destruction when the heavy monsoon rains begin to fall. In contrast, lightning fatalities in many industrialized nations are increasingly dominated by people involved in leisure activities in exposed locations rather than those engaged in outdoor occupational activities.

A tall tree should be avoided during a thunderstorm as its height is a very good 'attractor' of lightning. The stepped leader forming a cloud-to-ground lightning strike approaches the ground in steps or spurts (20 to 50 m /66 to 164 ft) until it makes contact with an upward streamer from the ground. An isolated 20- to 30-m-high (66 to 98 ft) tree is a strong contender to divert the final downward leader step, which then attaches to the tree to initiate lightning. The lightning earths via the tree trunk or side flashes during its descent of the trunk to a better electrical conductor nearby, which may be a person, animal or fence. Far too often, one of a group of people struck by lightning while sheltering under a tree may recall that someone in the group had commented that they ought not to be in such a location because of the increased risk it posed.

A person in a relatively open area who is holding a large metal or lengthy object such as a golf club, umbrella, fishing rod or spade is obviously much shorter than a mature tree. Even so, the upward streamers being generated from a person holding such an object provide a more preferential attachment point than the surrounding flat ground within, say, 2 or 3 m (7 or 10 ft). This increases the chances of being struck, but only if lightning was about to strike within that very small radius anyway. The implication is that people should avoid holding tall objects during thunderstorms, especially in open spaces.

A mountain is an exposed and dangerous location during a thunderstorm. Lightning near Deming, New Mexico.

The safest place to be during a thunderstorm is inside a large, substantive building that conducts the lightning around the outside, protecting its occupants. Some public awareness campaigns stress this advice: 'When thunder roars, go (stay) indoors', or, more graphically, 'Don't get fried, go inside!' Although a building is generally safe, especially if it has properly earthed wiring and plumbing circuits, there are occasions, as explained previously, when lightning strikes a building or nearby and sends a surging electric current along the electrical and plumbing circuits. Anyone who is in contact with the circuits or appliances (telephone, metal sink) linked to them may experience an electrical shock if lightning strikes the building.

Enclosed vehicles, with solid metal roofs and sides, are relatively safe refuges. This is because the metallic body acts as a Faraday cage and conducts the lightning safely around the occupants before earthing to the ground across the tyres, which are usually wet, or arcing the short distance from the metal bodywork to the ground. Occasionally, the lightning strike may damage the tyres, leave pitted marks on the vehicle's bodywork or shatter the windscreen. Occupants are advised to keep the windows closed and their arms away from the vehicle's sides to ensure the electrical current does not enter the interior or make contact with them. Open-top vehicles and soft-top convertibles are not safe. When lightning struck a truck in Texas in 1979, three passengers sitting in the open back were killed but the three inside the cab were uninjured.[33]

If caught out in the wide open with lightning striking close by and with no prospect of seeking protection in a substantive building or enclosed motor vehicle, there are precautions that can be taken to reduce the risk of being struck. You should move away from large or long metal objects such as wire fences, find a place of lower elevation such as a hollow, gully or dry ditch and adopt the 'lightning crouch or squat'. This means crouching as low as possible to reduce your height and tuck your head down – it is potentially less harmful to be struck on the shoulder than the head. Place your hands on your knees. Keep your feet together to reduce the ground current (step voltage effect) which may be

# When Thunder Roars Go Indoors!

*STOP Activities*

Seek shelter immediately in a substantial building or a hard-topped metal vehicle!

 **www.lightningsafety.noaa.gov**

Lightning seen through the windscreen of a car outside Winchester, UK. An enclosed vehicle, with solid metal roof and sides, is a relatively safe refuge when lightning threatens.

Poster of 'Leon the Lightning Lion' promoting lightning safety awareness for children.

experienced if lightning strikes nearby. This is also why you should not lie flat on the ground. This position is an uncomfortable one to maintain for long so is a last resort for someone caught out by lightning already striking near them.

## Wildfires

Wildfires initiated by lightning pose a threat to the lives of people, livestock and wildlife and cause considerable loss of natural vegetation and commercial forests. Towns and villages may be destroyed and cultural and historic sites may be lost forever. In countries such as Australia, Canada, China and the United States, lightning is a major cause of wildfires, a term which includes forest, grassland and bush fires. The other main cause, and one

Noticeboard warning hikers on the mountains of Sierra Nevada, to return to the safety of their vehicle if thunderstorms develop.

that accounts for an increasing number of fires, is people (whether they start the fires accidentally or deliberately). However, lightning-initiated wildfires are often the most extensive simply because they may occur in less accessible areas, making them difficult to tackle.

Some actions taken to lessen the wildfire hazard are forecasting the risk of lightning and lightning-initiated fires, early detection of fires by satellite and aircraft, and prescribed burning (the undertaking of controlled burning to reduce the amount of flammable material available to fuel a fire).

Hundreds of professional firefighters put their lives at risk when tackling a wildfire. In the United States lightning ignited a wildfire near Yarnell, Arizona, on 28 June 2013 and with severe drought conditions, dry air, high temperatures (38°c, 101°f) and strong winds, the fire spread rapidly and erratically. By 1 July the Yarnell Hill Fire had expanded to 8,300 acres (3,360 ha). Communities were evacuated as 400 firefighters struggled to control it. On 30 June thunderstorms had developed east of the fire, causing the winds to strengthen and the fire to flare up. It reversed direction and killed nineteen firefighters from an elite team known as the Granite Mountain Hotshots. Only one of the twenty-man team survived, as he was further away moving the team's trucks at the time. The firefighters had deployed individual fire protection shelters but not all of the bodies were found inside them. A fire shelter is a mound-shaped safety device constructed of layers of aluminium foil, woven silica and fibreglass. It is a last resort used by firefighters when trapped by wildfires, designed to reflect radiant heat, protect against convective heat and trap breathable air (most firefighters' deaths are caused by inhaling hot gases). In terms of loss of life, this was the deadliest wildfire of any kind in the United States since 1991, and the greatest loss of firefighters in the United States since the 11 September 2001 terrorist attacks in New York City. The wildfire was eventually contained by 10 July 2013. More than 3,000 people attended a public memorial service for the firefighters. The loss of life was high but the scale of the wildfire was small compared with many others. For example, from 9 May to 23 July the previous year, lightning strikes ignited two separate fires, which then combined and burned nearly 300,000 acres (121,400 ha) of the Gila National Forest in New Mexico. It took more than 1,200 firefighters to contain the blaze.

There are typically 100,000 wildfires each year in the United States, burning 4 million acres (1.6 million ha) to 9 million acres (3.6 million ha) of land. Lightning currently causes around 15

Lightning may start wildfires
in inaccessible areas making
them difficult to tackle quickly
and effectively.

Coconino National Forest
wildfire (Canyon Fire) in
Arizona on 2 July 2012
where 6,500 acres (2,600
ha) had already burned
after four days.

per cent of wildfires in the United States and 85 per cent is caused by people, either accidentally or intentionally, and other natural causes. However, because lightning initiates fires in less accessible areas, it accounts for around 60 per cent of the total area burned. If climate changes and cloud-to-ground lightning frequency increases, the problem may worsen significantly.[34]

British Colombia, the Yukon and the Northwest Territories suffer the worst wildfires in Canada, with around 9,000 wildfires each year. Currently around 35 per cent of the fires are initiated by lightning and the rest are started by people, most by accident. Fires started by lightning often occur in remote locations and in clusters and are typically about ten times as extensive as human-caused fires. The mean annual burned area from wildfires in Canada is 6 million acres (2.5 million ha) and 85 per cent of the total is caused by lightning.[35]

Australia's most damaging bushfire, known as the Black Saturday bushfires, began on 7 February 2009 in Victoria State. The fires occurred during extreme bushfire weather conditions with intense heat (temperatures reached 46°C or 115°F), little or no rain during the previous two months and winds exceeding 100 km/h (62 mph). Four hundred fires had been ignited, some by lightning but the majority by toppled or clashing power lines and by being lit deliberately. The fires resulted in Australia's highest ever loss of life from a bushfire with 173 people killed and more than 400 people injured. The fires covered 1,100,000 acres (450,000 ha) and affected 78 townships northeast of Melbourne. More than 7,500 people were displaced and the fires destroyed 2,000 houses and 3,500 structures, damaging thousands more. Nearly 12,000 sheep and cattle were killed. The estimated cost of the Black Saturday fires amounted to AUS$1 billion.[36] Australia suffers around 50,000 bushfires each year but most are relatively small.

## Farm animals

In addition to the threat to farms of lightning-initiated wildfires, lightning may strike, damage and set alight farm buildings and equipment as well as kill large numbers of farm animals. It is not only farm animals that are at risk from lightning strikes: highly valuable pedigree animals, racehorses, game birds and endangered zoo and safari park species as well as much-loved pets may be killed.[37]

Whole herds of animals may die in a single lightning incident. In China in July 2012 in three separate incidents, lightning struck and killed 173 sheep in the Xinjiang Uyghur autonomous region, 143 goats in Chagankule village, Xinjiang, and 53 pigs in the Guangdong province.[38] Historical accounts may far exceed these numbers. In Utah lightning killed 654 sheep on 18 July 1918 in the American Fork Canyon, while in the Raft River Mountains it killed 835 sheep and knocked a shepherd temporarily unconscious on 1 September 1939.[39] At Campshead Farm in Lanarkshire, Scotland, a lightning strike killed a farmer's son and 320 sheep in June 1748.[40]

Although a single lightning strike may kill and injure a large group of people, the number of animals killed may be much larger. This is because farm animals tend to bunch closely together, and they may shelter under a tree or crowd against a metal fence during a thunderstorm. One or more animals may be killed by the electrical current of a direct strike or a side flash (splash) from a nearby tree or fence. Others may die because the animals are touching and the deadly current passes from one to another. Even more may be killed by the ground current or step voltage effect; the latter is particularly dangerous for four-legged animals, whose legs span a greater distance than in humans. This produces a greater voltage gradient between the legs, which generates a larger electrical current up and down the legs of the animals and through their bodies. Unlike in humans, the route taken by this electrical current is more likely to pass through, or close to, life-sustaining organs. For some farmers the financial loss from the death of large numbers of animals is huge and made worse if the animals are uninsured, and if the carcasses cannot be sold because of traditional or religious beliefs that any animal killed by lightning should not be eaten.

Lightning striking farm areas may kill and injure large numbers of animals in a single strike.

## Buildings and structures

In thunderstorm-prone areas, incidents of lightning causing damage to buildings are frequent. Tall structures such as church towers, mosques and public buildings as well as high-rise office and apartment blocks experience more strikes than one- and two-storey houses, but usually suffer less damage. This is because these taller and often more costly buildings are likely to have lightning protection systems installed. Buildings fitted with such systems, providing they are well maintained, may be struck several or even many times a year with no resulting damage (and contradicting the myth that lightning never strikes twice in the same place).

Whereas public and commercial buildings are expected to implement national lightning protection standards, sometimes related to the local lightning ground flash density, the decision to fit lightning protection systems to one- and two-storey residential homes is usually left to the homeowner. Few have lightning protection systems fitted because lightning strikes are a low risk and need to be weighed against the cost of installing a system. Consequently, when lightning strikes such buildings, they may suffer serious damage to their structure and contents.

In the centuries before the lightning rod was invented, some incidents of lightning damage to buildings had been incredible. During the early hours of 27 October 1697, lightning struck Athlone Castle in Ireland, setting alight the contents of the munitions store, which exploded:

> It blew up 260 barrels of gunpowder, 1,000 charged hand-grenades, with 810 skains of match which were piled over them; 220 barrels of musket balls and pistol balls; great quantities of pick-axes, spades, shovels, horse-shoes and nails, all of which blew up into the air, and covered the whole town and neighbourhood and the fields for a great distance, by the violence of which the town gates were all thrown open. The poor inhabitants

Large tall buildings are likely to have lightning protection systems installed to conduct a lightning strike safely to earth. Lightning strikes a building in Denver, Colorado.

who were generally asleep when this tragical scene began, awakened with the different surprising misfortunes that befell them. Some finding themselves buried in the ruins of their own houses; others finding their houses in a flame over the heads; others blown from their beds into the streets; others having their brains knocked out with the fall of great stones, and breaking of hand-grenades in their houses.

The castle and town were destroyed, including the homes of 100 families, but, surprisingly, only seven people were killed and 36 wounded.[41]

Modern protection of buildings involves two components. First is the protection of the building's structure by intercepting the lightning current and diverting it safely into the ground. This may be achieved using one or more air terminals (lightning rods) installed on the roof or highest part of a building. They must extend at least 25 cm (10 in.) above the structure and are lightning strikes to various components of the city's electrical generation, transmission and distribution system left nine million people in New York City with no electricity for 24 hours. One factor in the slowness of restoring service was the need to manually reset the circuit breakers that had tripped during the collapse of the system. The cost of the blackout was estimated at around $350 million, with at least half this cost due to looting and arson. The city suffered outbreaks of violence and looting with 3,776 arrests made by the police and 1,037 fires reported by the fire department.[44]

Some protection of high-voltage electricity or power transmission lines is achieved by shielding. A wire is positioned above the transmission wires carrying the electricity supply in order to intercept any lightning strike in the vicinity. The protective wire is connected to the metal support towers and conducting wires enable the lightning current to flow safely down the tower into the ground to a buried grounding electrode.

Protection is also needed for the excessive voltages induced by lightning strikes. A short-lived voltage surge may travel along a transmission line so protective measures have to be fitted to divert the surge safely to the ground to prevent it reaching and damaging transformers (which convert the high voltage to low voltage electricity for use in homes and offices) and other equipment. Although lower voltage distribution supply lines are less well protected than the high-voltage transmission systems, they are less likely to be struck by lightning because they are much lower in height. Despite all the investment in protection systems, lightning is responsible for about 30 per cent of all power outages (interruption or failure of the power supply) in the United States at a total annual cost approaching $1 billion.[45]

## Ships and boats

Until the lightning rod was invented by Benjamin Franklin in 1752, the masts, sails and rigging of wooden ships were frequently damaged and set on fire, and crew and passengers seriously injured or killed. For example, the British frigate *Lowestoffe*, lacking any

Lightning causes
frequent power cuts
despite the huge sums
invested in protection.
Lightning strikes an
electricity substation
near Vaughan, New
Mexico.

lightning protection, was struck three times in quick succession on 8 March 1796 near Minorca, splintering and setting fire to the ship and killing or severely injuring the crew, many of whom were left with burns and paralysed limbs. In the 40 years from 1793 to 1832, more than 250 wooden ships of the British Navy were damaged by lightning and over 200 crew were killed or severely disabled. In the worst lightning incidents, ships were sunk with the loss of all those on board. If lightning struck a warship and the powder magazine exploded, the total loss of both ship and crew was likely.[46]

Ships and boats are particularly vulnerable to lightning strikes because they may be the only object that projects above sea level in a large area. The early form of lightning protection for wooden ships was to adapt Franklin's ideas. A short metal rod was placed at the masthead connected to long links of thin metal chain or long links of wire rope which were passed over the ship's side and dangled in the water. However, as the chains or wire rope were suspended from high up on the mast and then hung down to reach into the sea, they got in the way of changing sail or

connected to copper sheathing on the hull or to a galvanized iron earth plate on the underside of the hull in contact with the sea.[48] However, some of those on board small boats enjoying a fishing trip may sometimes forget that they remain vulnerable to a lightning strike if their fishing rod exceeds the height of the lightning rod.

Lightning striking a sailing warship around 1830, a decade before the permanent protection system invented by William Snow Harris became commonplace.

## Aircraft

Lightning costs airlines enormous sums each year, primarily not through the damage it causes to aircraft and equipment or death and injuries to passengers but through the huge costs invested in research and testing to ensure the safety of an aircraft and its passengers and crew. The routing of flights to avoid thunderstorms – avoiding potential lightning strikes as well as turbulence causing discomfort to passengers – causes delays and leads to extra fuel costs. Refuelling of aircraft on the ground has to be suspended if lightning threatens because of the high risk of fire and explosion, causing delay and disruption to flight schedules.

Commercial aircraft fly most days, except when they are undergoing maintenance, and are struck by lightning in flight on average only once or twice a year. This frequency contrasts markedly with that of the F-106 fighter aircraft used by the National Aeronautics and Space Administration (NASA) in an eight-year lightning research programme in the 1980s at the Langley Research Center in Hampton, Virginia. The instrumented F-106 had an all-metal exterior, relied on hydraulic rather than electronic controls, and was flown 1,496 times into thunderstorms to induce lightning. It succeeded 714 times. The research helped improve lightning protection standards for commercial and military aircraft and found that the aircraft triggered lightning strikes more often towards the top of a thunderstorm, particularly in the anvil. The F-106 pilots recollected: 'Our old rule was fly high and fry.'[49]

An aircraft may intercept a natural lightning strike in or near a thunderstorm, but more often the lightning is triggered by the aircraft itself flying through a region of high electrical charges, usually within clouds. In such cases, the lightning would not have happened if the aircraft had not been present to initiate it. Typically a lightning discharge will attach to the nose or wing tip, travel through the exterior skin of the fuselage, and exit safely off some other extremity, such as the tail. This envelops the aircraft in a large and changing magnetic field which, without adequate protection, would induce voltages in the lengthy wires threading through the aircraft, resulting in damage to electronic systems and computers. Certification requires that all aircraft must be rigorously tested and airframes proven to be able to withstand lightning strikes and that no on-board technical equipment, including fuel tanks, is at risk of damage in any way. Aircraft are designed to keep lightning on the outside of the plane and the only physical evidence of a lightning strike may be small burn marks on the extremities of the aircraft. As aircraft have become more complex, governments have upgraded certification requirements to require improved shielding and other protective measures to ensure electronics systems are unaffected.

With increasing reliance on computerized flight control systems, which control the aircraft in the safest and most efficient way, it is essential that the computer in this fly-by-wire system is unaffected by a lightning strike. The increasing use of composites (typically multilayered fine fibres made of carbon and glass held together by epoxy) instead of metal (such as aluminium) in the manufacture of aircraft has posed new challenges to ensure protection from lightning as they are less conductive than metal. For many years, the nose cone (radome) has been made of composite material to avoid interference with the radar lying inside it. To divert and safely conduct lightning away from the radome, thin metallic strips are incorporated onto its surface to stop the lightning puncturing the radome and damaging its electronics. One solution for the composite skin of an aircraft is to weave a thin layer of conductive fibres into the carbon fibres or add a thin aluminium or copper mesh. This adds weight but spreads the electrical currents of the lightning to minimize damage to the skin where lightning attaches, and keeps the current on the exterior of the aircraft. This effect helps reduce voltages that might be induced inside the aircraft, damaging electrical and electronic systems.[50]

Most lightning strikes to commercial aircraft simply result in the crew and passengers seeing a bright flash, hearing a loud bang, experiencing a temporary flickering of instruments and internal lights, and nothing else. Incidents of aircraft suffering temporary damage to the electronic systems, such as loss of flight information on the aircraft airspeed and altitude, are uncommon. Instances of more serious damage are rare and those resulting in loss of life are extremely rare.

Past aircraft disasters have galvanized international efforts to improve safety, such as on 8 December 1963 when lightning struck a Boeing 707 while it was flying in a holding pattern over Elkton, Maryland. The strike caused a spark that ignited fuel vapour in a tank. The resulting explosion resulted in the aircraft crashing, killing all 81 passengers. Half an hour after take-off from Lima Airport, Peru, on 24 December 1971, 91 passengers and crew of a Lockheed L-188A Electra were killed when lightning caused a fire on the right wing. The wing separated, causing the aircraft to crash into the Amazon rainforest. Remarkably, one seventeen-year-old woman survived the crash and lived for ten days in the remote, inhospitable rainforest before being rescued. On 8 February 1988 lightning struck a Fairchild Metro III commuter airliner as it approached Dusseldorf airport inside clouds at 900 m (3,000 ft). It was powered by two turboprop engines

Laboratory testing of an aircraft radome to ensure it is not damaged by lightning.

and carrying nineteen passengers and two crew members. The lightning resulted in the aircraft losing all electrical power and the pilot struggled to stabilize the aircraft. Eyewitnesses reported that the aircraft dived out of the cloud and then climbed back again several times. Eventually the landing gear, which had been deployed before lightning had struck, and a wing were torn from the aircraft. It went into a spiral dive and crashed with the loss of all on board.[51]

Pilots can be affected by momentary blindness from a lightning strike, especially at night (if cockpit lights are not at full brightness), but incidents in which a pilot experiences an electrical shock are exceptional. When lightning struck the windscreen of a Boeing 757 as it was approaching Amsterdam's Schiphol Airport at 1,500 m (5,000 ft) in October 2000, the first officer, who was at the controls, heard a loud bang, saw a bright blue flash and experienced a feeling he described as if he had been kicked in the chest. He recollects that he had brushed the windscreen just when it was struck. After recovering from the initial shock of the strike, he experienced difficulty in using his right arm, and passed control of the aircraft to the captain. Later medical examination found a chest wound consistent with an electrical discharge. No damage to the aircraft was subsequently found but the first officer, who returned to work after a fortnight, began to suffer from an irregular heartbeat that may have been a consequence of the incident.[52]

Lightning strikes the Space Shuttle Launch Complex 39A in the hours preceding the launch of *Challenger* on 30 August 1983.

## Spacecraft

The potential damage posed by lightning during the period leading up to, and during, the launching of spacecraft has caused particular concern. The potential for disaster was highlighted when the Saturn rocket carrying the *Apollo 12* command module triggered lightning during its launch through clouds on 14 November 1969, nearly aborting the second trip to land on the moon. Lightning had been triggered even though no 'natural' lightning was present. In March 1987 an unmanned Atlas-Centaur rocket triggered a lightning strike during its launch,

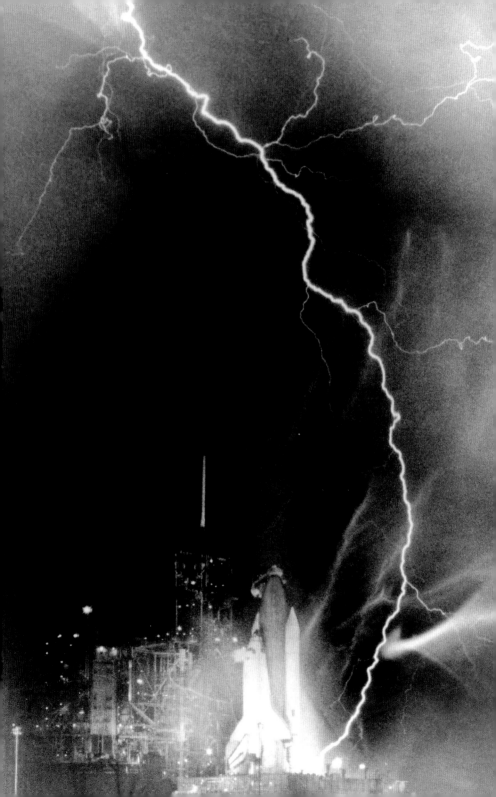

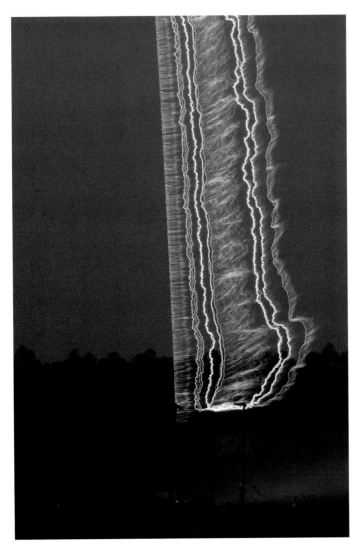

Lightning initiated by a small rocket trailing a copper wire at the ICLRT. It was triggered during tropical storm Debby in 2012 in the absence of natural lightning.

which punched through the fibreglass nose cone, damaged the computerized program instructions and sent the rocket veering off course 51 seconds after lift-off. NASA had no choice but to destroy the $160 million (at 1987 prices) rocket and naval communications satellite it was carrying.[53]

It seems that a rocket and its accompanying exhaust plume – aircraft do not have such long exhaust plumes – create an excellent electrical conducting route to the ground such that the threshold electrical field needed to trigger lightning is lowered. As a result of this increased threat, NASA and other national space agencies apply strict regulations to prevent launching when lightning may be triggered by the spacecraft as well as on occasions when thunderstorms are nearby to generate lightning anyway. One-third of all weather delays in the programme of U.S. Space Shuttle launches (1981–2011) were caused by the possibility that the launch vehicle and exhaust plume might have triggered lightning.[54]

The John F. Kennedy Space Center is in Florida, where frequent thunderstorms and lightning strikes occur. Between 5 per cent and 10 per cent of the annual 25 million cloud-to-ground strikes in the United States take place in Florida. As a result, a significant amount of lightning research is undertaken there including the deliberate triggering of cloud-to-ground lightning using small rockets. This research is being undertaken by the University of Florida at the International Center for Lightning Research and Testing (ICLRT) at Camp Blanding, Gainesville, established in 1993. Each rocket is 1 m (3 ft) long and trails a thin 700-m (2,300-ft) Kevlar-reinforced copper wire which unravels from a spool at the base of the rocket as it ascends into a thunderstorm. The wire is connected to a designated strike point on the ground. About half of the rocket launches trigger lightning. Lightning strikes the tip of the rocket and immediately vaporizes the trailing copper wire, leaving an ionized plasma channel in the air, which carries the current down to the grounding point. The lightning channel is unusually straight compared with natural strikes, although sometimes it deflects from the path, as defined by the wire, and strikes elsewhere. Clustered around the strike zone are a range of instruments and detectors that measure the size of the current, the power of the sound waves, the brilliance of the flash and the emission of x-rays. In recent years ten to 30 lightning discharges have been triggered by rockets each summer. Naturally occurring strikes within the grounds of the 100-acre (40-ha) research facility are studied too. Researchers test a variety of devices to determine their ability to

withstand a lightning strike, including insulation on aircraft and nuclear weapons, overhead power distribution lines, surge protectors, underground cables and gas pipelines, airport runway lights (as employed at small, unstaffed airports), a test house and golf course shelters. Research using rockets to trigger lightning have been, or are currently being, undertaken in Alabama and New Mexico, and in other countries including Brazil, China, France and Japan.[55]

Lightning protection at the John F. Kennedy Space Center has steadily evolved as the space programme has progressed. A combination of improved lightning protection masts and overhead wires has offered increased shielding of space vehicles. The protection system diverts electrical currents down to the ground that may have otherwise short-circuited electronics, including guidance and navigation systems, and damaged other equipment and payloads of scientific experiments. In 2009 lightning

Lightning protection system (three tall masts with connecting wires) at launch pad 39B, John F. Kennedy Space Center. The launch pad is shown being converted for a new generation of launch vehicles.

protection towers were installed at Kennedy Space Center's Launch Pad 39B to protect the next generation of space vehicle launches. The three 150-m (500-ft) steel towers, located around the pad perimeter, support 30-m (100-ft) fibreglass lightning masts and a network of wires stretching between the towers. Lightning striking the wires or towers will result in the surge of electrical current being diverted away from the launch vehicle and carried safely into the ground.[56]

## Is suppressing lightning a dream?

The worldwide costs of the damage, disruption and lives lost by lightning are enormous, as are the costs of investing in developing ways of minimizing these adverse impacts. Attempts to suppress or eliminate lightning in a locality have been made, but have yet to be shown to be effective. Lightning rods (air terminals) simply provide a safe passage for lightning to reach the ground. When the tip of the rod is fitted with arrays of multiple sharp metal points, brush-like in appearance, they are called static dissipation arrays. They are intended to reduce the electric field around the tip of the rod. Arguably, this may inhibit an intense upward streamer from forming, which may otherwise attach to the downward stepped leader and initiate lightning. The effectiveness of such systems remains the subject of ongoing research.

Attempts to eliminate or suppress lightning should be considered in the context of the tremendous power of a thunderstorm to generate a continual bombardment of thousands or even tens of thousands of lightning flashes. A proven way to initiate lightning is by firing a small rocket with a trailing wire into a thunderstorm. The use of a laser beam to trigger lightning is at an early stage of development but may be another means of doing so.[57] Whatever the means of triggering lightning, it is unlikely to make any significant or lasting diminution of the local electrification process that generates the lightning. Research conducted at the ICLRT concluded that the triggering of individual lightning flashes using small rockets did not significantly

Thunderstorms
produce lightning,
hail, damaging winds
and even tornadoes.
Suppressing any of
these powerful elements
has yet to be successful.

Dracula, killing a succession of young women and, in the guise of a bat, slaughtering all the women and children from the village near Dracula's castle who have sought shelter inside a church. He then captures a young man whose brother and girlfriend attempt his rescue. Dracula is eventually confronted on the ramparts of his castle but throws one of the rescuers over the turret, before another manages to plunge a metal spike into the vampire's chest. However, it doesn't strike his heart; Dracula pulls the stake out, raising it to impale the hero, but before he can do so, lightning suddenly strikes the spike and engulfs Dracula in flames and he falls, burning, from the castle's battlements.

Lightning occurs in films in many different ways. In the first full-length animated feature film, Walt Disney's *Snow White and the Seven Dwarfs* (1937), lightning is employed to save the dwarfs from a hopeless situation. After the evil Queen has poisoned

Lightning and thunder is a common backdrop in films to create fear or signify evil characters such as Dracula.

Snow White the dwarfs chase her into the mountains. There she attempts to dislodge a boulder to crush the dwarfs, but lightning strikes the ledge on which she is standing, causing her to fall and be crushed herself. Some films use the Greek and Roman storm god myths, such as *Percy Jackson and the Lightning Thief* (2010) and *Clash of the Titans* (1981, 2010). The comedy-drama film *Struck by Lightning* (2012) is the tale of a high school student suddenly struck and killed by lightning in a car park. In the highly successful science fiction film *Back to the Future* (1985), the scientist Doc Brown realizes that in order to send the hero, Marty McFly, back to the future, the power of a lightning bolt, which will strike the clock tower the following Saturday night at 10.04 p.m., must be harnessed by his time machine, the DeLorean. Intriguingly, in the American disaster movie *War of the Worlds* (2005) invading aliens are transported from their spaceship down lightning strikes into previously buried tripod machines, which then emerge from the ground.

Generally, in natural disaster movies, lightning tends to feature in a supporting role, because it is exceptional for a single lightning incident to cause more than one fatality or more than just localized damage. In a list of more than 130 disaster movies, lightning features as the main cause of destruction in only two – *Lightning: Fire from the Sky* (2001) and *Lightning: Bolts of Destruction* (2003). By contrast, hurricanes, tornadoes, volcanoes, earthquakes, tsunamis and meteors pose far more epic threats, attracting film-makers and audiences alike.[4]

In comic strip cartoons and subsequent film adaptations, several characters have embraced lightning in their name to signify their supernatural power, speed and invincibility as well as their ability to use lightning as a powerful weapon. Black Lightning, introduced in 1977 as the alter ego of Metropolis-based Olympic decathlete Jefferson Pierce, was one of the first African-American superheroes to appear in DC Comics. He generates electrical energy as a flash of lightning and can move at the speed of light. He believes that 'Justice, like lightning, should ever appear to some men hope, to other men fear.' Black Lightning had two daughters who, as costumed superheroes,

*Tornado*, engraving by William Blake after a drawing by Henry Fuseli, 1795. Lightning is shown as a bunch of zigzag lightning flashes.

and shows a yellow zigzag 'lightning bolt' passing from top left to bottom right, becoming 'an Art Deco equivalent of a comic book explosion'.[10]

In paintings lightning is usually bright white, but in reality it may appear in many colours as the shorter wavelengths of light (blue, indigo, violet) are scattered more strongly by air molecules than light at longer wavelengths (red, orange). If lightning is relatively near, most of the light at all wavelengths reaches the viewer and the lightning appears brilliant white. If the lightning is distant, the blues and greens are often depleted by the intervening air, leaving the reds, oranges and yellows more prominent. The same phenomenon occurs with the sun, which appears dazzling white when high in the sky, but frequently red-orange when setting. Natural dust, particulate pollutants and water droplets can scatter wavelengths of light too so, for example, lightning may appear yellow in colour in a dust-laden sky. In the United States, forest rangers consider white lightning more likely to start fires than red lightning, because red light is released when the hydrogen atoms present in a moist atmosphere are 'excited' by the intense heat of lightning. By contrast, white lightning indicates an absence of rain which could dampen a fire. Icy blue lightning is associated with hail. During thundersnow, pink and green lightning have been observed.

An engraving of zigzag lightning striking and sinking a French warship, St Lawrence River, Canada, in the 18th century. All aboard were drowned, with the exception of two women.

Fulgurites or 'petrified lightning tubes' have featured in the work of artist Allan McCollum, famed for his extensive conceptual art installations. He collaborated with the University of Florida's ICLRT to create fulgurites artificially in 1998. A small rocket, with a rapidly unravelling thin copper wire attached to it, was fired into a thunderstorm. Lightning was triggered through the wire, which was connected to a container on the ground

Allan McCollum
revealing fulgurite
created by triggering
lightning using a
rocket-and-wire.
A fulgurite formed in
the container of sand
into which the lightning
was directed.

filled with dry compacted sand. McCollum experimented with
various types of sands, with different melting points and grain
shapes. The best result came with a tube of pure zircon sand,
which produced a slender, pale grey fulgurite. McCollum used
this as a mould to make 10,000 duplicates which he then displayed side by side on felt-covered tables for visitors to reflect
on the merging of art and science.[11]

The fulgurites created by McCollum closely resembled
naturally produced fulgurites and greatly contrast to the film
portrayal of a fulgurite found on a beach in the American roman-
tic comedy, *Sweet Home Alabama* (2002). The lead male character,
Jake, creates a fulgurite by placing a metal pole in the sandy beach
during a thunderstorm to channel the lightning into the ground.
When lightning strikes the pole and fuses the sand at its base to
create a fulgurite the result is shown to be a beautiful, crystal clear
and smooth-blown glass sculpture with antler-like extensions.

Some of the 28 tables displaying 10,000 fulgurites.

The original fulgurite from which 10,000 duplicates were created for Allan McCollum's display *The Event: Petrified Lightning from Central Florida (with Supplementary Didactics).*

Unfortunately it bears no resemblance to a real fulgurite and was created by a company which specializes in making stunning glass sculptures.

The largest sculpture associated with lightning is *The Lightning Field* created by the American sculptor Walter De Maria (1935–2013) in 1977. This enormous work of art is situated in a remote area of the high desert of western New Mexico. It consists of 400 polished stainless steel poles – lightning rods – installed in a grid array measuring 25 poles (each 5 cm or 2 in. in diameter) long (1,600 m or one mile) by 16 poles wide (1 km or 3,200 ft); that is, the poles are each just over 60 m (200 ft) apart. It takes about two hours to walk around the sculpture. The land is undulating but the poles are of varying height, from 8 m (15 to 26 ft), so the solid pointed tips align in a horizontal plane. The sculpture is dramatic to view at sunrise and sunset in the desert when the metal rods glow in the red, pink or golden light but is spectacular about 60 times a year, when it is struck by lightning.[12]

## Mimicking thunder

Percussion instruments were devised in the nineteenth century to enable Western classical orchestras to mimic thunder. A thunder sheet is a suspended sheet of metal, up to around 5 m (16 ft) long, which can be shaken or hit with a drumstick to produce a rumble of thunder. German composer Hans Werner Henze's (1926–2012) Sixth Symphony, written in 1969, uses two thunder sheets, one small and one large. Some organs have a 'storm-effects' stop which sounds two low-frequency pipes that, when combined, imitate a storm. Other attempts to mimic thunder have included pebbles poured into a large metal container, bags of stones thrown against a metal surface, lead balls dropped onto a sheet of leather, and heavy stones or balls of lead rolled down a slatted ramp. The musical score for *An Alpine Symphony*, op. 64, written in 1915 by German composer Richard Strauss (1864–1949), employed a thunder machine (*Donnermaschine*) consisting of a rotating drum with balls inside.[13] Other musical scores have simply employed drum

Beautiful, crystal-clear glass sculptures were wrongly claimed to have been created fulgurites in the film *Sweet Home Alabama* (2002).

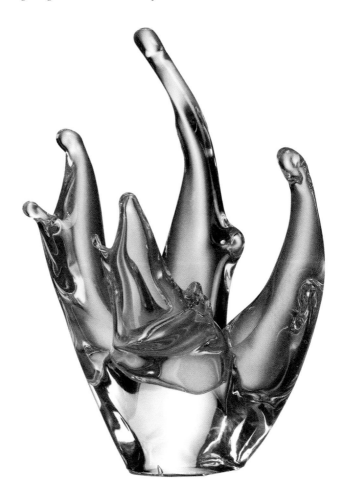

rolls and cymbal clashes to simulate thunder, including the 1868 polka *Unter Donner und Blitz* ('Thunder and Lightning'), op. 324 by Austrian composer Johann Strauss (1825–1899). Frequent use of clashing cymbals conveys the thunder and lightning in the title track of the Irish hard rock band Thin Lizzy in their final studio album *Thunder and Lightning* (1983). Strobe lighting effects and backdrop film of lightning flashes added impact to their live performances of this track.

An early attempt to mimic thunder in the theatre gave rise to the common accusation 'stealing [one's] thunder', when it was first coined by John Dennis (1657–1734), an English critic and playwright. In 1709 he staged the play *Appius and Virginia*, in which he claimed he had created a new way to simulate the sound of thunder. The method is not known but may have involved metal balls rolled around in a wooden bowl. When the play was performed at the Drury Lane Theatre, London, it failed to attract sufficient audiences and so the theatre cancelled it. When Dennis returned to the theatre to watch a performance of Shakespeare's *Macbeth*, he realized they were using the same thunder technique. He stood up and shouted 'That's my thunder, by God! The villains will not play my play but they steal my thunder.'[14]

## Iconography

The zigzag symbol has often been depicted on the tepees and clothes of Native American tribes and was commonly painted on the faces of warriors as part of their war paint. Although the meaning may vary from one tribe to another, the symbol – associated with the legendary Thunderbird – was believed to add power and speed to the warrior. The importance of regular rain for growing food in the dry American Southwest, including New Mexico, explains why the zigzag lightning symbol features in some Pueblo Indian petroglyphs too.

The Pueblo Indians also represent lightning in drawings of snakes. Not only are snakes considered to represent lightning in their shape but they were thought to be able to generate lightning. Consequently live snakes featured in rainmaking ceremonies, which were first recorded by outsiders in the early 1890s. As part of the nine days of Snake Ceremonials performed by Hopi Indians, a sub-tribe of the Pueblo Indians who live in northern Arizona, a large sand painting was created on the floor of an underground prayer room or kiva. The sand mosaic depicted a mass of clouds from which emerged four coloured lightning streaks (yellow, green, red and white) in the form of snakes. The four snake images corresponded to the four cardinal points of the compass and the four quarters of the world. Two of the snakes were identified as male by adding a curved horn, and the two female with the addition of a square with diagonals. Numerous short black lines were drawn outside the frame on a background of white sand to represent rain. As many as 100 live snakes, including deadly rattlesnakes, were consecrated by washing them in charm liquid. Accompanied by an increasingly louder and wilder song and drumming, punctuated by fierce, bloodcurdling yells or war cries, the snakes were thrown with great force onto a sand painting where they writhed, becoming covered in the different coloured sands. The ceremony is intended to invoke lightning and, more importantly, the rain that accompanies a thunderstorm. During the ceremonies, the priests wore clothing on which were woven zigzag snake/lightning bands and they carried small clay balls on which they had marked zigzag lines with their thumbnails. Dancers carried a wooden zigzag-shaped 'lightning staff' representing a snake. After the ceremony, the priests took

A sand mosaic made by Pueblo Indians in Arizona, used in traditional rainmaking ceremonies. Different coloured snakes, symbolizing lightning, are shown emerging from storm clouds.

pinches of sand from each of the different coloured clouds and lightning symbols and carried them to the fields. The traumatized snakes were returned to the four cardinal points of the plains and foothills from where they had been captured. They were being sent as messengers to the Hopi gods and the spirits of their ancestors in the hope that they would answer their fervent prayers for rain.[15]

Most commonly today, the lightning zigzag is employed internationally as an electrical symbol. It is used on some vehicle dashboard and machinery controls to indicate an electrical fault and is displayed on signs to indicate the potential danger of an electrical hazard, such as a high-voltage electrical transformer, electrical machinery or equipment, overhead or buried electrical cables and electric fences. This symbol even has an International Organization for Standardization (ISO) number, namely ISO 3864. It is sometimes made even more explicit in signs by depicting an electrical spark striking a person. A bespoke sign may be used to warn a specific group of vulnerable people that there is a nearby electrical hazard, such as those fishing where overhead electrical cables are present.

The zigzag symbol, with its association with electricity, was used in a large public sculpture in Sydney, Australia, to raise awareness of high levels of domestic energy consumption and to encourage energy conservation. A 6.5-m (21-ft) zigzag-shaped sculpture, *Thunderbolt*, was made from recycled metals. It was commissioned in 2010 from Bonita Ely to celebrate the tenth anniversary of the Olympic Games held in Sydney. The

Wooden lightning staff (1913) in the shape of a snake carried by male Pueblo Indian dancers during rainmaking ceremonies in Arizona. The snake in a zigzag shape is believed to be magically linked to lightning.

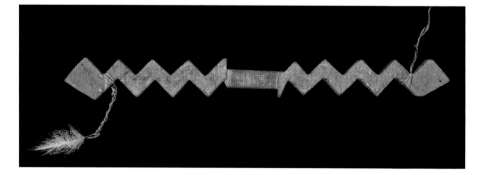

The zigzag-shaped lightning symbol is an internationally recognized electrical symbol, displayed where there is potential danger of an electric shock.

local community's nocturnal energy use was monitored through Energy Australia and live data were sent to the sculpture to control the colours of its low-level lighting system, powered by solar panels. The illumination of the 'environmental signal' sculpture changed from green (low energy consumption) to yellow (tipping point) to red (high energy consumption). In this way, it aimed to encourage a change in consumer behaviour in the surrounding community. After eighteen months at the Sydney Olympic Park, the sculpture was relocated for permanent display at Broken Hill, New South Wales.

The lightning zigzag shape or 'thunderbolt' features in several installations by Bonita Ely, including the 6-m (20-ft) tubular steel *Lake Thunder*, which overlooks Thuy Tien Lake at Hue, Vietnam. The sculpture glows in the dark 'as a sceptre of the earth's energies at night'. The setting for this sculpture is intended to reflect traditional Taoist philosophical principals about the 'four forms' inherent in us: 'thunder' for our true essence, 'lake' for our true sense, 'water' for our real knowledge and 'fire' for our conscious knowledge.[16]

Given that the zigzag lightning symbol represents an electrical symbol and lightning is associated with power, speed, swiftness and even invincibility, it has become a common insignia for military communications. A bundle of lightning flashes may occasionally be adopted, such as with the 2nd Signal Regiment of the Spanish Army. In flags, shields and heraldic coats of arms for individuals and a range of organizations distinctive zigzag symbols may be incorporated, such as for the city of Imatra in Finland.

*Thunderbolt* sculpture by Bonita Ely, Sydney Olympic Park, Australia, 2010. The 6.5-m (21-ft) sculpture, or 'environmental signal', changes colour at night in response to the amount of electricity consumed in the neighbourhood. The green colour indicates low consumption.

*Lake Thunder* steel sculpture, 6 m (20 ft), overlooking Thuy Tien Lake, Hue, Vietnam, by Bonita Ely, 2006.

The lightning symbol has been adopted as a distinctive motif by some sports teams and political parties. Since they were established in 1992, Florida's professional ice hockey team have been known as Tampa Bay Lightning or, simply, the Bolts. Their shirts display a blue or white lightning zigzag logo that reflects in part the long-held belief that the Tampa Bay area is the 'lightning capital of North America', even if this is based simply on the average number of days of thunderstorm activity annually rather than more precise measures of lightning activity. The People's Action Party (PAP) in Singapore, founded in 1954, adopted a readily recognizable red lightning flash superimposed on a white background and blue circle. PAP explains the lightning flash stands for 'action' in its pursuit of a just and equal society.

In the past, the zigzag symbol appears to have been embraced by several far right political parties, such as the National Socialist Movement of Chile, which existed from 1932 to 1938. It initially supported facism and the ideas of Adolf Hilter, receiving financial support from Chilean people of Germanic descent. The now

Lightning symbols
in 'coats of arms':
2nd Signal Regiment,
Spanish Army (left)
and City of Imatra,
Finland (right).

defunct British Union of Fascists adopted the lightning flash,
too. The latter party formed in 1932 but was banned in 1940 by
the government; the party's leader, Oswald Mosley, and many
hundreds of its members were interned for much of the Second
World War. The National States Rights Party (NSRP), a far right
party, incorporated a red lightning flash. It was founded in 1958
in Knoxville, Tennessee, and opposed the American Civil Rights
Movement, preaching racism and anti-Semitism. The double
zigzag emblem was also adopted by the Schutzstaffel (ss), the
paramilitary or secret police of the Nazi Party from 1933 to 1945.
The abbreviation ss was depicted on flags and uniforms by two
Sig runes drawn side by side, looking like lightning bolts. The
original Sig rune represented the sun, not lightning, in the Old
Norse and Old English runic alphabets, but it was reinterpreted
by the Nazi Party, with its interests in Germanic mysticism, to
symbolize their rallying cry 'Victory, Victory'. The Schutzstaffel,
led by Heinrich Himmler, was responsible for many crimes
against humanity during the Second World War and, along with
the Nazi Party, was banned in Germany in 1945.[17]

Tarot cards can be traced back to their first use in Italy in the
fifteenth century, and today the lightning zigzag symbol features

'The Tower' or 'Lightning-struck Tower' Tarot card, which may signify an unexpected catastrophe for the person whose fortune is being read.

Stunning lightning – nature's own spectacular firework display – above a hotel lobby, at Friona, Texas.

on one of the most unwelcome Tarot cards. It is depicted on the sixteenth of the major group of Tarot 22 cards within the Rider-Waite-Smith deck, the most common deck used in the English-speaking world, with intricate designs used in fortune telling. The sixteenth card is named 'The Tower' or, more fully, 'The Lightning-struck Tower'. It shows a tower being struck by lightning, the top of the tower (shown as a crown) toppling and two people falling headfirst from the tower, presumably to their death. The appearance of 'The Tower' card in someone's spread usually means that a sudden, drastic and unforeseen change is imminent, and the lightning implies that the cause of this is a force beyond their control. The card is unwelcome for the unexpected catastrophe and turmoil it brings to what they had thought was a safe, secure situation (the inside of the tower). One positive interpretation is that it brings new beginnings – although not by choice. However, the symbolism in Tarot cards is very complex and the meaning of a card is modified according to its position in the spread of cards, the meaning of adjacent cards and whether it is upside down.[18]

For many children as well as adults in recent decades, the lightning zigzag symbol is associated with the scar on the forehead of the young wizard Harry Potter in J. K. Rowling's book series. The lightning-shaped scar is the result of a failed murder attempt by the evil wizard Lord Voldemort on 31 October 1981 when he struck fifteen-month-old Harry with a 'Killing Curse', which Harry survived because his mother sacrificed herself to protect him. In many countries Halloween (Hallowe'en or All Hallows' Eve) is celebrated on 31 October each year by children wearing scary costumes of witches, vampires, ghosts and skeletons. Since the *Harry Potter* books were published, wearing a Harry Potter costume, brandishing a wand and drawing the lightning symbol on the forehead is another popular guise for children as they go from house to house 'trick-or-treating' to receive treats or perform mischief.

## Final reflections

As one of nature's most powerful elements, lightning has engendered the widest range of emotions in humans throughout their existence. Early civilizations initially depended on lightning to ignite fires from which they could collect a burning branch to light their hearths and give them the warmth and the ability to cook food and boil water to survive. At the same time, they cowered with fear at the dangers it posed – their leaders and shamans gave them storm gods to whom they could pray for protection from the storm's wrath and, in many cases, for these gods to address their other needs too, such as ensuring adequate rain for crops. While we no longer welcome the fires it ignites and, instead, try to limit the number of wildfires and buildings it sets alight, lightning continues to create fear in some people and anxiety in many others. This may encourage some to ask whether or not we can control and prevent lightning. However, as discussed, the effective suppression or elimination of lightning must remain, for the time being, simply a dream for the future.

Nevertheless, because of the challenges that lightning poses to us in relation to the threat to people, wildlife, environment, buildings, transportation, electricity supplies and in many other areas of our lives, we need to undertake more research to improve our knowledge of lightning and how we can limit and cope with its negative impacts. This means continuing to undertake scientific studies, make technical advances and adjust our lives and activities to cope with the lightning risk. Ensuring everyone understands the types of locations and activities that place them at risk from lightning is a priority for educational and public awareness campaigns. Changing people's attitudes is critical so that outdoor activities which may put people at risk are scheduled to take place at a safe time by referring to weather forecasts. Equally, if such activities are taking place when thunderstorms are developing nearby then they need to be curtailed promptly in order for participants to seek safe refuge.

The feeling of fear or anxiety of lightning among some societies is matched by respect and, in many others, perhaps even excitement. Watching a thunderstorm from a safe location as it generates thunderous displays of lightning is awe-inspiring. We see each flickering flash as unique in the pathway it follows and the form it takes. It is unpredictable and fickle. Each cloud-to-ground lightning flash may surprise us when it happens, as may the precise location of the cloud from which it emerges, and where it strikes the ground. Differences in brightness and colour may astonish us too. The spectacle of lightning is nature's beauty at its finest.

It is not surprising that 'capturing lightning', albeit with a camera, has become a goal for many people. Some of these are 'storm chasers' who seek close encounters with nature's extremes, especially in the American Midwest. Although they are usually more interested in intercepting the paths of severe thunderstorms to photograph any tornadoes that may be spawned, photographs of lightning spectacles may be a bonus for them. For other photographers in that region and in other locations around the world, capturing and sharing images of the stunning beauty and diversity of nature's 'fireworks' is their primary aim. For the rest of us,

Lightning in Dumas, Texas.

Lightning at
night in Texas.

at least we can enjoy looking at the photographs of lightning taken by these individuals without facing any of the dangers inherent to the chase.[19] Appreciating the stunning beauty of lightning must never make us forget that lightning can be deadly and dangerous.

# GODS AND GODDESSES OF THUNDER AND LIGHTNING

### Africa

*Lightning Bird (Umpundulo, Ndlati)*, southern Africa
*Set (Seth)*, Egypt
*Shango (Sango) and Oya*, Yorùbá, West Africa

### Americas

*Chaac (Chak), God K (K'awil) and God N (Pauahtun)*, Maya-Toltec, Mesoamerica
*Changó and Olla*, Cuba
*Ilyap'a (Illapa, Catequil, Apocatequil)*, Incas
*Kadlu (and her sisters Kweetoo and Ignirtoq)*, Inuits
*Thunderbird and Lightning Snake*, Native Americans
*Tlaloc*, Aztecs, Mesoamerica
*Xangô and Uansa*, Brazil

### Asia

*Adad (sometimes Addu)*, Akkadians, Mesopotamia
*Anatolia*, Hittites, Mesopotamia
*Hadad (Haddu)*, Arameans, Mesopotamia
*Indra*, Vedic Hinduism, India
*Ishkur (Iškur)*, Sumerians, Mesopotamia
*Ishtar*, Babylonians, Mesopotamia
*Lei Gong (Lei Shen)*, Lei Tsu and Dian Mu (or Tien Mu), China
*Raijin (Raiden)*, Japan
*Rammon (Rimmon)*, Assyrians, Mesopotamia
*Teshub*, Hittites, Mesopotamia

**Europe**

*Donar*, Germany
*Horagalles*, Sami (Lapps), northern Scandinavia
*Jupiter and Summanus (for nocturnal lightning)*, Romans
*Perkons*, Latvia
*Perkunas*, Lithuania
*Perun*, Russia and Ukraine
*Pikker*, Estonia
*Piorun*, Poland
*Taranis*, Celtic regions such as Gaul
*Thor (Thunor)*, Norse, Scandinavia
*Tinis (Tin or Tinia)*, Etruscans
*Ukko*, Finland
*Zeus and Athena (his daughter)*, Greece

**Oceania (including Australasia)**

*Lightning Brothers (Yagdabula and Jabiringi)*, Aborigines, Australia
*Lightning Man (Namarrgon, Namarrkun, Mamaragan)*, Aborigines,
    Australia
*Pele (Pélé) and kane-hekili*, Polynesia and Hawaii
*Tawhaki and Whaitiri*, Maoris, New Zealand

13  Martin A. Uman, *The Art and Science of Lightning Protection* (Cambridge and New York, 2008), pp. 212–14.

14  For studies on fulgurites, see Regina A. Lee, 'Fulgurites: "Paleo–Lightning" Remnants', *Meteorology 4990* (4 December 1992), www.usfcam.usf.edu, accessed 21 August 2013; George P. Merrill, 'Fulgurites, or Lightning-holes', *Popular Science Monthly*, XXX (1998); and Julian J. Petty, 'The Origin and Occurrence of Fulgurites in the Atlantic Coastal Plain', *American Journal of Science*, Fifth Series, XXXI/183 (1939), www.usfcam.usf.edu, accessed 21 August 2013.

15  Sibley, *The Divine Thunderbolt*, p. 31.

16  Andrew Dickson White, *A History of the Warfare of Science with Theology in Christendom* (New York, 1898), chapter XI: '"The Prince of the Power of the Air"', section III: 'The Agency of Witches'.

17  'Witch Trials in the Early Modern Period', http://en.wikipedia.org, accessed 25 September 2013.

18  Folger Shakespeare Library, '*Daemonologie*, in the Form of a Dialogue, Divided into Three Books', www.folger.edu, accessed 26 September 2013.

19  White, *A History of the Warfare of Science*, p. 363.

20  See Wolfgang Behringer, *Witches and Witch-hunts: A Global History* (Cambridge and Malden, 2004).

21  Estelle Trengove and Ian Jandrell, 'Lightning and Witchcraft in Southern Africa', paper presented to 7th Asia-Pacific International Conference on Lightning, November, Chengdu, China (2011), pp. 173–7.

22  B. L. Meel, 'Witchcraft in Transkei Region of South Africa: Case Report', *African Health Sciences*, IX (2009), pp. 61–4.

23  'Case Study: The European Witch-hunts *c.* 1450–1750 and Witch-hunts Today', www.gendercide.org, accessed 21 April 2013.

24  'Witchcraft in Asia and sub-Saharan Africa', www.religioustolerance.org, accessed 6 March 2013.

25  'Congo Incident: Lightning Kills Football Team', http://news.bbc.co.uk, 28 October 1998; and 'South African Incident: Footballers Struck by Lightning', http://news.bbc.co.uk, 26 October 1998.

26  Houston Chronicle, 'Villages Hound "Witches" after Lightning', http://wwrn.org, 2 January 2003.

27  On Benedict Daswa see 'The Price of Being Christian', www.benedictdaswa.org.za, accessed 23 February 2015.

28  Vatican Radio, 'Benedict Daswa to be Beatified as Pope Francis Gives his Assent', http://en.radiovaticana.va, accessed 23 February 2015.

29  Trengove and Jandrell, 'Lightning and Witchcraft'; Adrian Koopman, 'Lightning Birds and Thunder Trees', *Natalia*, XXXIXI (2011), pp. 40–60.

30  White, *A History of the Warfare of Science with Theology in Christendom* (New York, 1898), chapter XI: 'From "The Prince of the Power of the Air" to Meteorology', section II: 'Diabolical Agency of Storms'.

31  Ibid., p. 345.

32  Ibid., pp. 346–7.

33  Mojca Kovačič, 'The Bell and its Symbolic Role in Slovenia', ICTM Study Group on Folk Musical Instruments: Proceedings of the 16th International Meeting (2006), pp. 108–9, www.llti.lt, accessed 23 April 2013.

34  Ron Hipschman, 'Lightning: The Whys and Wherefores of Nature's Fireworks', www.exploratorium.edu, accessed 24 September 2013.

35  White, *A History of the Warfare of Science*, p. 334.

36  'Agnus Dei: Lamb of God', www.rosaryworkshop.com, accessed 24 September 2013.

37  Sergio Bertelli, *The King's Body: Sacred Rituals of Power in Medieval and Early Modern Europe* (University Park, PA, 2001), pp. 135–6.

38  Reginald Scot, *The Discoverie of Witchcraft* (1584), Book XII, chapter IX, www.esotericarchives.com, accessed 25 September 2013.

39  Bertelli, *The King's Body*, p. 136.

40  The historical basis is explored by Harry F. Williams, 'Old French Lives of Saint Barbara', *Proceedings of the American Philosophical Society*, CXIX/2 (1975), pp. 156–85.

41  Christian Bouquegneau and Vladimir Rakov, *How Dangerous is Lightning?* (New York, 2010), pp. 9–10.

42  Sibley, *The Divine Thunderbolt*, pp. 285–8; John S. Friedman, *Out of the Blue: A History of Lightning* (New York, 2009), pp. 32–6.

43  Estelle Trengove and Ian Jandrell, 'Strategies for Understanding Lightning Myths and Beliefs', *International Journal of Research and Reviews in Applied Sciences*, VII (2011), pp. 287–94.

44  Abbas, 'Houseleeks (*Sempervivum tectorum*): Superstitions, History and Medical Benefits', 26 November 2010, http://herbs-treatand-taste.blogspot.co.uk, accessed 24 August 2013; 'Berlin's Weather Talisman', *The Observer* (30 June 1907), p. 5.

45  Eugene S. McCartney, 'Why did Tiberius Wear Laurel in the Form of a Crown during Thunderstorms?', *Classical Philology*, XXIV (1929), pp. 201–3.

46  Catharine Edwards, *Suetonius: Lives of the Caesars. A New Translation* (Oxford, 2000), sections 58 and 89; James Hastings and John A. Selbie, *Encyclopedia of Religion and Ethics*, II (1910), p. 51, https://archive.org, accessed 2 October 2013.

47  Deane P. Lewis, 'Owls in Mythology and Culture' (2005), www.owlpages.com, accessed 2 October 2013.

48 James George Frazer, 'Some Popular Superstitions of the Ancients, Folk-lore', *Quarterly Review of Myth, Tradition, Institution and Custom*, I (1890), pp. 153–4.

49 Trengove and Jandrell, 'Strategies for Understanding Lightning Myths and Beliefs'.

## 3 The Science and Nature of Lightning

1 Early estimates of worldwide thunderstorm frequency were provided by C.E.P. Brookes, 'The Distribution of Thunderstorms Over the Globe', *Geophysical Memoirs*, III/24 (1925), pp. 147–64. He suggested that there were 1,800 thunderstorms at any one time from which was inferred a lightning frequency of 100 lightning flashes per second. However, his often quoted data were taken from land-based reports of thunderstorms and assumed the storms were as frequent over the sea as over land, which is not the case. Modern detection provides much lower figures; Hugh J. Christian et al., 'Global Frequency and Distribution of Lightning as Observed from Space by the Optical Transient Detector', *Journal of Geophysical Research*, CVIII/4005 (2003), pp. 1–15; National Oceanic and Atmospheric Administration, 'Science on a Sphere: Annual Lightning Flash Rate Map', http://sos.noaa.gov, accessed 6 October 2013.

2 Vladimir A. Rakov and Martin A. Uman, *Lightning: Physics and Effects* (New York, 2003), pp. 9–10.

3 Christian et al., 'Global Frequency'; 'World Lightning Map', http://geology.com, accessed 16 September 2013; National Aeronautics and Space Administration (NASA), 'NASA Science News: Where Lightning Strikes', http://science.nasa.gov, accessed 15 October 2013.

4 Colin Price and David Rind, 'Possible Implications of Global Climate Change on Global Lightning Distributions and Frequencies', *Journal of Geophysical Research*, IC (1994), pp. 10,823–31; Colin Price, 'Thunderstorms, Lightning and Climate Change', paper presented at the 29th International Conference on Lightning Protection, 23–6 June, Uppsala, Sweden (2008), www.iclp-centre.org, accessed 26 June 2013; American Friends of Tel Aviv University, 'Climate Change May Lead to Fewer but More Violent Thunderstorms', *Science Daily*, www.aftau.org, 10 July 2012; David M. Romps et al., 'Projected Increase in Lightning Strikes in the United States due to Global Warming', *Science*, CCXLVI (2014), pp. 851–4.

5 For general characteristics of lightning, see Martin A. Uman, *Lightning* (New York, 1984); Rakov and Uman, *Lightning: Physics*

*and Effects*; and Ivan Amato, 'Dark Lightning: The Unseen Energy of Thunderstorms', www.guardian.co.uk, 23 April 2013.

6 Derek Elsom, 'Learn to Live with Lightning', *New Scientist*, CXXII/1670 (1989), pp. 54–8; Clive P. R. Saunders, 'A Review of Thunderstorm Electrification Processes', *Journal of Applied Meteorology*, XXXII (1993), pp. 642–55.

7 Martin A. Uman, *All About Lightning* (New York, 1986), p. 76; Peter E. Viemeister, *The Lightning Book* (New York, 1961), pp. 59–61, 111–15.

8 Uman, *Lightning*, pp. 5–10, 14–46; John S. Friedman, *Out of the Blue: A History of Lightning: Science, Superstition, and Amazing Stories of Survival* (New York, 2008), pp. 110–14.

9 Dan Robinson, 'Upward-moving Lightning from TV Towers, Skyscrapers and Other Tall Structures', http://stormhighway.com, accessed 14 September 2013.

10 Uman, *Lightning*, pp. 181–201; Arthur A. Few, 'Thunder', in *Atmospheric Phenomena* (San Francisco, CA, 1975), pp. 111–21.

11 Mark Twain, Letters of Mark Twain, VI, 1907–1910. Quote is from a letter written on 28 August 1908, www.gutenberg.org, accessed 15 October 2013.

12 John Bostock and H. T. Riley, *The Natural History of Pliny the Elder* (London, 1855), I, p. 70.

13 Friedman, *Out of the Blue*, p. 38.

14 Ibid, pp. 77–8; Viemeister, *The Lightning Book*, pp. 31–3; 'Science and Curiosities at the Court of Versailles: The First Electricity Experiment in the Hall of Mirrors', http://en.chateauversailles.fr, accessed 15 October 2013.

15 Friedman, *Out of the Blue*, pp. 78–82; Viemeister, *The Lightning Book*, pp. 35–43; E. Philip Krider, 'Benjamin Franklin and Lightning Rods', *Physics Today* (January 2006), pp. 42–6; The Franklin Institute, 'Ben Franklin's Lightning Bells', www.fi.edu, accessed 16 October 2013.

16 Benjamin Franklin and William Temple Franklin, *Memoirs of the Life and Writings of Benjamin Franklin*, vol. III (Philadelphia, PA, 1808), p. 113.

17 Krider, 'Benjamin Franklin'; 'Revolutionary War and Beyond: Benjamin Franklin and Electricity', www.revolutionary-war-and-beyond.com, accessed 16 October 2013.

18 Andrew Dickson White, *A History of the Warfare of Science with Theology in Christendom* (New York, 1898), chapter XI: 'From "The Prince of the Power of the Air" to Meteorology', section IV: 'Franklin's Lightning Rod'.

19 'Pliny the Younger's Observations', http://home.utah.edu, accessed 16 October 2013.

20 MIT Technology Review, 'How Sandstorms Generate Spectacular Displays', www.technologyreview.com, accessed 17 October 2013.
21 Friedman, *Out of the Blue*, pp. 85–6.
22 On ball lightning see Mark Stenhoff, *Ball Lightning: An Unsolved Problem in Atmospheric Physics* (New York, 1999); Amarendra Swarup, 'Physicists Create Great Balls of Fire', *New Scientist*, www.newscientist.com, 7 June 2006; Roger J. Jennison, 'Ball Lightning', *Nature*, CCXXIV/895 (1969).


**4 Lightning Threats to People and Activities**

1 Ronald L. Holle, 'Annual Rates of Lightning Fatalities by Country', paper presented at the 20th International Lightning Detection Conference, 24–5 April (Tucson, AZ, 2008), www.vaisala.com, accessed 6 October 2013; Ronald L. Holle, 'Some Aspects of Global Lightning Impacts', paper presented at the 7th Conference on the Meteorological Applications of Lightning Data, 4–8 January (Phoenix, AZ, 2015), www.ams.confex.com, accessed 20 February 2015. Kenneth Wilson, 'Death, Injury by Lightning Strike in Cambodia can be Reduced', *Cambodia Daily*, www.cambodiadaily.com, 18 March 2013.
2 Derek M. Elsom, 'Deaths and Injuries Caused by Lightning in the United Kingdom: Analyses of Two Databases', *Atmospheric Research*, LVI (2001), pp. 325–34; Ronald L. Holle, 'Recent Studies of Lightning Safety and Demographics', paper presented at the 22nd International Lightning Detection Conference, 2–3 April (Broomfield, CO, 2012), www.vaisala.com, accessed 6 October 2013; Derek M. Elsom and Jonathan D. C. Webb, 'Lightning Deaths and Injuries in the United Kingdom, 1988–2012', *Weather*, LXIX (2014), pp. 221–6; Derek M. Elsom, 'Deaths Caused by Lightning in England and Wales, 1852–1990', *Weather*, XLVIII (1993), pp. 83–90; Derek M. Elsom, 'Striking Reduction in the Annual Number of Lightning Fatalities in the United Kingdom since the 1850s', *Weather*, LXX (2015), in press.
3 'The Damp Stick', www.youtube.com, accessed 25 September 2013.
4 E. Philip Krider, 'Benjamin Franklin and Lightning Rods', *Physics Today* (2006), pp. 42–6, www.physicstoday.org, accessed 16 October 2013.
5 'Swedish Footballer Struck by Lightning is Back on the Ball', www.independent.co.uk, 10 August 1994.
6 M. A. Cohen, 'Clinical Pearls: Struck by Lightning', *Academic Emergency Medicine*, VIII (2001), pp. 893, 929–31; Mary Ann Cooper, Christopher J. Andrews and Ronald L. Holle, 'Lightning Injuries', in *Wilderness Medicine*, ed. Paul S. Auerbach, 5th edn (Philadelphia,

PA, 2007), pp. 67–108, www.uic.edu, accessed 21 March 2013;
M. O'Keefe Gatewood and R. D. Zane, 'Lightning Injuries',
*Emergency Medicine Clinics of North America,* XXII (2004), pp. 369–403.

7 Michael Cherington et al., 'Lichtenberg Figures and Lightning:
Case Reports and Review of the Literature', *Cutis: Cutaneous
Medicine for the Practitioner,* LXXX (2007), pp. 141–3; Cohen,
'Clinical Pearls'; A. L. Mahajan, R. Rajan and P. J. Regan,
'Lichtenberg Figures: Cutaneous Manifestation of Phone
Electrocution from Lightning', *Journal of Plastic, Reconstructive
and Aesthetic Surgery,* LXI (2008), pp. 111–13; T. Ocak et al., 'Two
Cases of Lightning Strikes Resulting in Lichtenberg Figures',
*Dermatologica Sinica,* XXX (2013), pp. 1–12.

8 Cooper et al., 'Lightning Injuries'.

9 Gretel Ehrlich, *A Match to the Heart* (London, 1995).

10 Lightning Strike and Electric Shock Survivors International, Inc.,
www.lightning-strike.org, accessed 24 November 2013.

11 Dionysius Lardner, *Popular Lectures on Science and Art; Delivered
in the Cities and Towns of the United States* (New York, 1859), II,
15th edn, p. 100, http://scans.library.utoronto.ca, accessed
20 October 2013.

12 Ibid., p. 101.

13 Derek Elsom, 'Learn to Live with Lightning', *New Scientist,*
CXXII/1670 (1989), pp. 54–8.

14 E. J. Heffernan, P. L. Munk and L. J. Louis, 'Thunderstorms and
iPods: Not a Good Idea', *New England Journal of Medicine,* CCCLVII
(2007), pp. 198–9.

15 S. Esprit, P. Kothari and R. Dhillon, 'Injury from Lightning Strike
while Using Mobile Phone', *British Medical Journal,* CCCXXXII/7556
(2006), p. 1513; C. W. Althaus, 'Injury from Lightning Strike while
Using Mobile Phone: Mobile Phones are Not Lightning Strike
Risk', *British Medical Journal,* CCCXXXIII/7558 (2006), p. 96;
R. M. Faragher, 'Injury from Lightning Strike while Using Mobile
Phone: Statistics and Physics Do Not Suggest a Link', *British
Medical Journal,* CCCXXXIII/7558 (2006), p. 96.

16 Cooper et al., 'Lightning Injuries'; Elsom and Webb, 'Lightning
Deaths and Injuries'; R. H. Golde and W. R. Lee, 'Death by
Lightning', *Proceedings of Institution of Electrical Engineers,*
CXXIII/10 (1976), pp. 1163–80; J. Gookin, 'Backcountry Lightning
Risk Management', paper presented at the 21st International
Lightning Detection Conference, 19–20 April (Orlando, FL, 2010),
www.pcta.org, accessed 2 October 2013; Vladimir A. Rakov and
Martin A. Uman, *Lightning: Physics and Effects* (New York, 2003);
Martin A. Uman, *The Art and Science of Lightning Protection*
(Cambridge and New York, 2008), pp. 125–9.

17 'Entire Football Team Struck by Lightning', www.thelocal.de, 9 August 2008.

18 Uman, *The Art and Science of Lightning Protection*, pp. 95–6, 127–9.

19 Mary Ann Cooper, 'A Fifth Mechanism of Lightning Injury', *Academic Emergency Medicine*, IX (2002), pp. 172–4.

20 'Rhys Jones Hit by a Series of Electric Shocks', www.dailyecho.co.uk, 22 December 2011.

21 Michael Cherington, Howard Wachtel and Philip R. Yarnell, 'Could Lightning Injuries be Magnetically Induced?', *The Lancet*, CCCLI (1998), p. 1788.

22 Elsom and Webb, 'Lightning Deaths and Injuries'.

23 Christopher J. Andrews and M. Darveniza, 'Telephone-mediated Lightning Injury: An Australian Survey', *Journal of Trauma*, XXIX (1989), pp. 665–71; Uman, *The Art and Science of Lightning Protection*, p. 216; S. Dinakaran, S. P. Desai and D. M. Elsom, 'Telephone-mediated Lightning Injury Causing Cataract', *Injury*, XXIX (1998), pp. 645–6; Mahajan et al., 'Lichtenberg Figures'.

24 'Alexander Mandon Buried Alive as "Cure" After Being Struck by Lightning Four Times', www.huffingtonpost.com, 27 March 2013.

25 Paul Thompson, 'Man Survives Being Struck by Lightning SIX Times', www.dailymail.co.uk, 30 June 2011.

26 John S. Friedman, *Out of the Blue: A History of Lightning* (New York, 2009), pp. 11–16.

27 Charlie Connelly, *Bring Me Sunshine* (London, 2012), pp. 59–61.

28 Dan Robinson, 'Lightning Safety: The Myths and the Basics', http://stormhighway.com, accessed 2 October 2013.

29 W. P. Roeder, 'Lightning Safety Procedures for the Public', paper presented at the 21st International Lightning Detection Conference, 19–20 April (Orlando, FL, 2010), www.vaisala.com, accessed 24 September 2013.

30 National Athletic Trainers' Association, 'Position Statement: Lightning Safety for Athletes and Recreation', *Journal of Athletic Training*, XLVIII (2013), pp. 258–70.

31 Evan Williams, 'Dodging Lightning: Schools and Parks Install Warning Systems to Prevent Tragedies', http://fortmyers.floridaweekly.com, 13 July 2013.

32 John S. Jensenius, Jr., 'A Detailed Analysis of Recent Lightning Deaths in the United States', www.lightningsafety.noaa.gov, accessed 10 October 2013.

33 Elsom, 'Learn to Live with Lightning'.

34 'Yarnell Hill fire', www.azcentral.com, accessed 12 October 2013; Burned Area Emergency Response (BAER) team executive summary, 'Whitewater-Baldy Complex (Gila National Forest), June 18, 2012', www.fs.usda.gov, accessed 10 October 2013; 'Wildfires',

http://environment.nationalgeographic.co.uk, accessed 25 September 2013; 'The Age of Western Wildfires', www.climatecentral.org, accessed 25 September 2013; 'U.S. Wildfires: Burn Area Expected to Double by 2050', www.huffingtonpost.com, 17 December 2012.

35 'Forest Fire', www.thecanadianencyclopedia.com, accessed 15 September 2013.

36 'Australia Scorched by 50,000 Bushfires a Year: Report', www. watoday.com.au, 26 November 2009; 'Black Saturday Bushfires', http://en.wikipedia.org, accessed 25 September 2013.

37 Chandima Gomes, 'Lightning Safety of Animals', *International Journal of Biometeorology*, LVI (2012), pp. 1011–23.

38 '173 Sheep Killed by Lightning', www.chinadaily.com.cn, 13 July 2012; 'Lightning Strike Kills 143 Goats in Xinjiang', www.china.org, 12 July 2012; 'Lightning Strikes and Instantly Kills 53 Pigs', www.chinahush.com, 9 July 2012.

39 'Death by Lightning for Giraffes, Elephants, Sheep and Cows', http://scienceblogs.com, accessed 29 September 2013.

40 Ross Finlay, *Touring Scotland: The Lowlands* (Henley-on-Thames, 1969), p. 82.

41 Anonymous, 'The Lightning Strike Which Blew Up the Magazine of Athlone Castle', *Journal of Meteorology*, XI/112 (1986), p. 272.

42 New York Press Office, 'Lightning Claim Costs Continue to Rise: Surging Electronics Prices, Product Shortages, Partly to Blame', www.iii.org, 12 June 2012.

43 'The 1999 Southern Brazil Blackout', http://en.wikipedia.org, accessed 14 October 2013.

44 R. Sugarman, 'Power/energy: New York City's Blackout – A $350 million Drain', *IEEE Spectrum*, XV (1978), pp. 44–6.

45 Uman, *The Art and Science of Lightning Protection*, pp. 25, 199–215; Smith, *Lightning*, pp. 185–95.

46 R. Anderson, *Lightning Conductors – Their History, Nature and Mode of Application* (London, 1879); 'Ships Struck By Lightning', *The Times* (1 March 1858) p. 9.

47 Hugh Cannell, *Lightning Strikes: How Ships are Protected from Lightning* (Brighton, 2011); T. Bernstein and T. S. Reynolds, 'Protecting the Royal Navy from Lightning: William Snow Harris and his Struggle with the British Admiralty for Fixed Lightning Conductors', *IEEE: Transactions on Education*, XXI (1978), pp. 7–14; Uman, *The Art and Science of Lightning Protection*, pp. 175–9.

48 Cannell, *Lightning Strikes*, pp. 167–76; Uman, *The Art and Science of Lightning Protection*, pp. 179–85.

49 'NASA Feature: Lightning Research', www.nasa.gov, 11 December 2007.

50 Jack Williams, 'How Things Work: Lightning Protection',
www.airspacemag.com, 1 July 2011.
51 Uman, *The Art and Science of Lightning Protection*, pp. 152–74;
Martin A. Uman and Vladimir A. Rakov, 'The Interaction of
Lightning with Airborne Vehicles', *Progress in Aerospace Sciences*,
XXXIX (2003), pp. 61–81; Juliane Koepcke, *When I Fell from the Sky:
The True Story of One Woman's Miraculous Survival* (London and
Boston, MA, 2012).
52 'Lightning Strikes BA Pilot in Mid-flight, Burns Hole in his Chest',
http://rense.com, accessed 15 June 2013.
53 Eliot Marshall, 'Lightning Strikes Twice at NASA', *Science*, CCXXXVI
(1987), p. 9; Uman and Rakov, 'The Interaction of Lightning'.
54 'NASA Educator Features: Lightning and Launches', www.nasa.gov,
22 April 2004.
55 Vladimir A. Rakov, 'A Review of Lightning-triggered Experiments',
paper presented at the 30th International Conference on Lightning
Protection (Cagliari, 2010), www.iclp-centre.org, accessed 20 October
2013; Vladimir A. Rakov, M. A. Uman and K. J. Rambo, 'A Review
of Ten Years of Triggered-lightning Experiments at Camp Blanding,
Florida', *Atmospheric Research*, LXXVI (2005), pp. 503–17.
56 'Lightning Protection for the Next Generation Spacecraft',
www.nasa.gov, 5 December 2007; 'Launch Pad 39B Boasts
Comprehensive Weather System', www.nasa.gov, 3 May 2011.
57 Jeff Hecht, 'Lightning Directed by Laser Beams', *New Scientist*,
2858 (30 March 2012).
58 Uman, *The Art and Science of Lightning Protection*, pp. 221–9.

**5 Lightning in Literature, Art and Popular Culture**

1 Academy of American Poets, 'Poets: Randall Jarrell',
www.poets.org, accessed 14 November 2013.
2 Mississippi Writers and Musicians, 'William Hodding Carter, II',
www.mswritersandmusicians.com, accessed 14 November 2013.
3 IEEE Global History Network, 'Galvani and the Frankenstein
story', www.ieeeghn.org, accessed 17 November 2013; Marco
Piccolini, 'Animal Electricity and the Birth of Electrophysiology:
The Legacy of Luigo Galvani', *Brain Research Bulletin*, XLVI (1998),
pp. 381–407.
4 Mike Sean, 'Acts of God: Natural Disaster Films' (2011),
www.imdb.com, accessed 15 November 2013.
5 DC Comics Database, 'Black Lightning', http://dc.wikia.com,
accessed 15 November 2013.
6 DC Comics Database, 'Captain Marvel', http://dc.wikia.com,
accessed 15 November 2013.

7   DC Comics Database, 'The Flash (Wally West)',
    http://dc.wikia.com, accessed 15 November 2013; Comic Vine,
    'Lightning Lad', www.comicvine.com, accessed 15 November 2013.

8   Jonathan Smith, *Charles Darwin and Victorian Visual Culture*
    (Cambridge, 2009), pp. 267–8; 'Stonehenge', www.victorianweb.org,
    accessed 22 November 2013.

9   'Roy Lichtenstein Thunderbolt', www.lamodern.com, accessed
    22 November 2013.

10  'Roy Lichtenstein: A Retrospective, Modern 1966–71',
    www.artic.edu, accessed 22 November 2013.

11  Jade Dellinger, 'Allan McCollum: Artist Interviewed',
    www.usfcam.usf.edu, accessed 21 August 2013; Adrienne
    M. Golub, 'Bolts from the Blue', *Art Papers*, XXIII/1 (1999),
    http://allanmccollum.net, accessed 21 August 2013; To supplement
    his exhibition, Allan McCollum made 66 published papers and
    interviews about fulgurites available at www.usfcam.usf.edu,
    accessed 21 August 2013.

12  Dia Art Foundation, 'Walter De Maria: The Lightning Field',
    www.diaart.org, accessed 15 November 2013.

13  K. L. Aplin and D. Williams, 'Meteorological Phenomena in
    Western Classical Orchestral Music', *Weather*, LXVI (2011),
    pp. 300–305; J. Blades, *Percussion Instruments and Their History*
    (London, 1970), p. 464.

14  'John Dennis (dramatist)', http://en.wikipedia.org, accessed
    15 November 2013.

15  J. Walter Fewkes, 'The Snake Ceremonials at Walpi', *Journal of
    American Ethnology and Archaeology*, IV (1894), pp. 13–126.
    www.bhporter.com, accessed 25 October 2013; Aby M. Warburg,
    *Images from the Region of the Pueblo Indians of North America*
    (Ithaca, NY, and London, 1997), pp. 3, 9–11, 36–7.

16  'Curating Cities: A Database of Eco Public Art, Thunderbolt –
    Bonita Ely', http://eco-publicart.org, accessed 25 October 2013;
    House of Laudanum, 'Thunderbolt', http://houseoflaudanum.com,
    accessed 25 October 2013.

17  'Runic Symbol of the Schutzstaffel, http://en.wikipedia.org,
    accessed 14 November 2013.

18  Arthur Edward Waite, *The Pictorial Key to the Tarot* (London, 1911).

19  Some of the websites which provide stunning images of lightning
    are listed in 'Associations and Websites' at the end of the book.

# SELECT BIBLIOGRAPHY

Andrews, Christopher, et al., *Lightning Injuries: Electrical, Medical and Legal Aspects* (Boca Raton, FL, 1992)

Blinkenberg, Christopher, *The Thunderweapon in Religion and Folklore* (Cambridge, 1911)

Bouquegneau, Christian, and Vladimir Rakov, *How Dangerous is Lightning?* (New York, 2010)

Cade, C. Maxwell, and Delphine Davis, *The Taming of the Thunderbolts: The Science and Superstition of Ball Lightning* (London and New York, 1969)

Cannell, Hugh, *Lightning Strikes: How Ships are Protected from Lightning* (Brighton, 2011)

Ehrlich, Gretel, *A Match to the Heart* (New York, 1995)

Friedman, John S., *Out of the Blue: A History of Lightning – Science, Superstition, and Amazing Stories of Survival* (New York, 2008)

Graf, Mike, *Lightning! And Thunderstorms* (New York, 1998)

Kithil, Richard, *Fundamentals of Lightning Protection: An Overview* (Louisville, KY, 2012)

——, *Introduction to Lightning Safety and Risk Management of the Hazard* (Louisville, KY, 2010)

Rakov, Vladimir A., and Martin A. Uman, *Lightning: Physics and Effects* (New York, 2003)

Renner, Jeff, *Lightning Strikes: Staying Safe under Stormy Skies* (Seattle, 2002)

Schonland, Basil, *The Flight of the Thunderbolts* (Oxford, 1964)

Sibley, Jane T., *The Divine Thunderbolt: Missile of the Gods* (Milton Keynes, 2009)

Smith, Craig B., *Lightning: Fire from the Sky* (Newport Beach, 2008)

Staller, John E., and Brian Stross, *Lightning in the Andes and Mesoamerica* (Oxford and New York, 2013)

Stenhoff, Mark, *Ball Lightning: An Unsolved Problem in Atmospheric Physics* (New York, 1999)

Uman, Martin A., *All About Lightning* (New York, 1986). This edition
     is an unabridged and corrected publication of Martin Uman's
     *Understanding Lightning* (Carnegie, PA, 1971)
——, *The Art and Science of Lightning Protection* (Cambridge and New York,
     2008)
——, *Lightning* (New York, 1984)
Viemeister, Peter E., *The Lightning Book* (New York, 1961)
White, Andrew Dickson, *A History of the Warfare of Science with Theology
     in Christendom* (New York, 1898)

# ASSOCIATIONS AND WEBSITES

Environment Canada – Lightning Safety
   http://ec.gc.ca

Lightning photographs
   http://all-that-is-interesting.com/incredible-lightning-photographs
   http://environment.nationalgeographic.com/environment
   http://chaseday.com
   www.lightningphotography.com
   www.thunderstorm-chaser.com
   www.weatherstudios.com

Lightning Strike and Electric Shock Survivors International, Inc.
   www.lightning-strike.org

Medscape – lightning injuries
   http://emedicine.medscape.com

National lightning location detection websites
   http://webflash.ess.washington.edu
   www.netweather.tv
   www.earthnetworks.com
   www.vaisala.com
   www.weatherzone.com.au

National Lightning Safety Institute (USA)
   www.lightningsafety.com

National Weather Service (USA) – Lightning Safety
   www.lightningsafety.noaa.gov

Struckbylightning.org – lightning strike database
   www.struckbylightning.org

The Royal Society for the Prevention of Accidents (UK)
– Lightning at leisure
www.rospa.com

Tornado and Storm Research Organisation (UK) – Lightning impacts
www.torro.org.uk

University of Florida – Lightning Research Group and the
International Center for Lightning Research and Testing
www.lightning.ece.ufl.edu

UK Meteorological Office
www.metoffice.gov.uk

Weather Wiz Kids
www.weatherwizkids.com

# ACKNOWLEDGEMENTS

This book developed from my lifetime interest in the weather, especially thunderstorms and the lightning, hail, strong winds and occasional tornadoes they generate. The opportunity to research and write a book devoted to lightning came, with the support of Daniel Allen and Michael Leaman of Reaktion Books, when I retired from Oxford Brookes University two years ago.

Apart from the fascination of watching lightning, one of the earliest influences on my desire to understand this weather phenomenon better came when I read Martin A. Uman's book entitled *Understanding Lightning* (1971), subsequently reissued as *All About Lightning* (1986). His ability to explain so many aspects about lightning in such a straightforward way, without using too many technical terms, set me the goal of one day writing my own reader-friendly book about lightning. Many years have passed since I first read Martin's book but I hope I have produced an accessible, interesting book about lightning which will prompt readers, especially those new to the subject, to want to find out even more about this amazing weather phenomenon.

My interests in lightning have been supported by colleagues in the Tornado and Storm Research Organization (TORRO), which I headed from 1994 to 2008. Many hundreds of TORRO members have collected and researched information for inclusion in UK databases of lightning incidents affecting people, animals, buildings, power supplies and transport. Jonathan Webb has played a key role in this. My research publications to assess the risk of injury and death to people from lightning have used this information. Other TORRO colleagues to thank are Peter van Doorn, Adrian James, Paul Knightley, Terence Meaden, Keith Mortimore, Robert Doe, Helen Rossington, Michael Rowe, Mark Stenhoff and John Tyrrell.

Many people have generously provided lightning photographs from which to select images for inclusion in this book. They include Kim Anderson, Jane Burridge, Chris Cameron-Wilton, Matthew Clark, Cammie Czuchnicki and Tim Moxon of Weather Studios, Nathan Edwards, Steve Hodanish, Glyn Jones, Steve Kay, Howard Kirby, Peter

Scott and especially John C. Wright. Several of these are members of TORRO or UKweatherworld (UKWW) and have undertaken annual visits to the American Midwest in their quest to capture stunning lightning images.

My appreciation goes to many individuals and members of organizations who have provided information and identified illustrations for me to use. They include Laura Baxter and Kathy McFall (National Health Service, Greater Glasgow and Clyde), Christopher Chatfield, Jeremy Coote (Pitt Rivers Museum, University of Oxford), Ashley Cummings (United States Department of Agriculture Forest Service), Paul Domaille, Bonita Ely (College of Fine Arts, University of New South Wales), Peter Faris (Rock Art Blog), Don Fuller (Contemporary Art Museum, University of South Florida), Sigurlaug Gunnlaugsdóttir and Thórður Arason (Icelandic Meteorological Office), Steven Haigh (Lightning Testing and Consultancy, Cobham Technical Services), Rachel Hunt (Total Lightning Network, Earth Networks), Peter Keene (Thematic Trails), Justin Kerr (Maya Vase Database), Richard Kithil (National Lightning Safety Institute), Martin Legault (National Resources Canada), Meghan Mahoney (Simon Pearce Glassware), Michael Nickell and Malcolm McElvaney (Sibley Nature Center, Midland, Texas), Leena Peura-Kuusela (City of Imatra, Finland), Dustin Hill, John Pilkey and Martin Uman (International Center for Lightning Research and Testing, University of Florida), Piper Severance (Foundation for the Advancement of Mesoamerican Studies), Katie Sneed (Lightning Protection Systems, LBA Group) and Linda and Heather Watson from Glasgow.

Estelle Trengove and Ian Jandrell (University of the Witwatersrand, Johannesburg) provided me with useful insights into the lightning myths and beliefs in South Africa. Their research aims at improving lightning safety among traditional communities. John E. Staller, an archaeologist and the world's leading authority on the significance of lightning for ancient religious ideologies, folklore and legend in Mesoamerica, was particularly helpful in suggesting appropriate images of storm deities.

Finally, I am grateful for the support received from friends, Richard Huggins and Jane Iliffe (Oxford Brookes University), and family: Sally, Matt and Nola Lacey, Clare Elsom (who, in her profession as an illustrator, produced three of the illustrations), Dan Maxwell, and my wife, Elizabeth.

# PHOTO ACKNOWLEDGEMENTS

The author and the publishers wish to express their thanks to the below sources of illustrative material and /or permission to reproduce it.

Kim F. Anderson: p. 185; Jane Burridge: p. 168; Ryan Bushby (HighinBC): p. 34; Matthew Clark: pp. 102–3, 210; Daderot: p. 33; El Caballero: p. 105 centre; Earth Networks: pp. 142, 143; Nathan Edwards: pp. 95, 163, 206–7; Bonita Ely: p. 203; Bonita Ely and Mr Snow (http://houseoflaudenum. com/art-technology/thunderbolt/): p. 202; Elsevier license number 3317130704944: p. 131; Derek Elsom: pp. 16, 35, 55 bottom, 74, 201; Derek Elsom/Clare Elsom: pp. 15, 137, 199; C. Goodwin: p. 52; Hans Hillewaert: p. 69; Steve Hodanish: p. 96; James Humphreys/Salopian James: p. 178; Icelandic Meteorological Office: p. 118 (Thórður Arason); Georges Jansoone JoJan: p. 70; Jschmeling: p. 38; Steve Kay: p. 164; Peter Keene: p. 152; Howard Kirby: pp. 100–101; Library of Congress, Washington, DC: p. 63; Lightning Protection Systems, LBA Group: p. 182; Lightning Testing and Consultancy, Cobham Technical Services: p. 173; M. G. Loppe/Camille Flammarion: p. 106; Los Angeles County Museum of Art (LACMA), contract #021402: p. 39; Marie-Lan-Nguyen: p. 20; Mary Evans Picture Library: pp. 19 (Classicstock/SIPLEY), 30, 31, 112, 115 (Classic Stock/H. Armstrong Roberts), 134, 170, 188 (Classic Stock/H. Armstrong Roberts), 193, 205; Malcolm McElvaney: pp. 55 top, 57; Michael McQuilken: p. 139; National Health Service, Greater Glasgow and Clyde, and Wilson family: p. 130; Simon Pearce Glassware Company: p. 197; Photograph K521, Justin Kerr: p. 36; Pitt Rivers Museum, University of Oxford: pp. 49 PRM 1901.49.1 to 1901.49.4, 200 PRM 1913.87.87; Science Photo Library: p. 175 (Science Source); Pete Scott: pp. 6, 98–9, 105 top, 126–7, 160–61, 180–81; Sibley Nature Center, Midland, Texas: p. 56 (Bill Loos); Spanish Armed Forces, and the City Imatra: p. 204; John E. Staller: p. 40; Estelle Trengrove: pp. 66, 77; United States National Weather Service/National Aeronautics and Space Administration (NASA): pp. 87; United States National Weather Service/National Oceanic and Atmospheric Administration (NOAA): pp. 145, 150; University of Florida: p. 54 (Martin A. Uman); University of Florida

*Photo Acknowledgements*

Lightning Research Group: p. 176 (Dustin Hill); University of South Florida Contemporary Art Museum: pp. 194, 195; USDA Forest service, Southwestern Region: pp. 119, 156–7; Weather Studios: pp. 86, 104, 107, 211 (Cammie Czuchnicki and Tim Moxon); Greg Willis: p. 28; Chris Cameron-Wilton: p. 151; John C. Wright: pp. 10–11, 22, 45, 80, 84–5, 88–9, 92–3, 94, 97, 116–17, 128, 144, 148, 154–5, 167. For the Michael McQuilken photographs on p. 139, permission thanks go to Michael McQuilken, 1975. The photograph cannot be published or posted anywhere without written consent from Michael McQuilken.

# INDEX